Synthesis Lectures on Biomedical Engineering

This series consists of concise books on advanced and state-of-the-art topics that span the field of biomedical engineering. Each Lecture covers the fundamental principles in a unified manner, develops underlying concepts needed for sequential material, and progresses to more advanced topics and design. The authors selected to write the Lectures are leading experts on the subject who have extensive background in theory, application, and design. The series is designed to meet the demands of the 21st century technology and the rapid advancements in the all-encompassing field of biomedical engineering.

Khalid Sayood

Biomedical Signal Processing

A One Semester Course

Springer

Khalid Sayood
Department of Electrical and Computer
Engineering
University of Nebraska—Lincoln
Lincoln, NE, USA

Nebraska R & D
Lincoln, USA

ISSN 1930-0328 ISSN 1930-0336 (electronic)
Synthesis Lectures on Biomedical Engineering
ISBN 978-3-032-08854-3 ISBN 978-3-032-08855-0 (eBook)
https://doi.org/10.1007/978-3-032-08855-0

This Springer imprint is published by the registered company Springer Nature Switzerland AG
The registered company address is: Gewerbestrasse 11, 6330 Cham, Switzerland

To Füsun

Preface

A signal is anything that carries information. When this something is measurable, we can process it, and it becomes a subject for signal processing. That is an enormous range of signals. To develop practical techniques, we need to narrow our focus. We can discuss speech processing, which enables us to handle a set of signals that can be characterized in ways that facilitate the development of speech processing algorithms. We can then broaden our scope and discuss audio signal processing. There is considerable continuity there, but even this relatively slight broadening of scope brings in different problems and opportunities. Similarly, we can examine image processing, which deals with signals that are very different from speech and audio signals and therefore requires a distinctly different approach. Broadening that to video processing brings us almost into an entirely different realm. In each of these applications, a commonality can be exploited: speech signals occupy a relatively consistent range of frequencies, regardless of the speaker. When processing images for human viewing, we must always be cautious of any distortions in the edges. Video signals are generally heavily correlated from one frame to the next. The signals we encounter when we enter the realm of biomedical signals are not as constrained. An EEG signal differs significantly from an EKG signal, which in turn differs from an ultrasound signal, which are all completely different from bioinformatic signals.

The fact that they are all biomedical signals does not constrain them in the same way as speech, audio, or video signals. This makes teaching a course in biomedical signal processing both interesting and challenging. Interesting due to the wide range of topics available, but challenging because of the lack of convenient resources. One of those convenient resources is a textbook. There are very nice expositions available for specific applications, such as EEG or EKG. However, when I began teaching the subject, I couldn't find a book that exactly fit how I wanted to teach the class (that might say something about me, but let's not go there), or the subjects I wanted to cover, and which could be used for self-study. Hence, this book.

The book consists of three parts: the first part covers basic signal processing concepts, the second part examines signals generated by the human body, and the third part explores signals used to study the human body. I assume the reader is familiar with the Fourier transform. I have included an appendix on Fourier transforms in case the assumption is incorrect.

With topics as diverse as those covered in this book, I needed the help of a large number of people to ensure I stayed on track, and I am very grateful to all of them. Mike Hoffman, my former colleague at the University of Nebraska and my current retirement friend, has always had a significant role in all my writing exercises—from books to exams. This particular project owes even more to him. Mike created the biomedical signal processing class, which I later took over, enjoying the fruits of his labors without having to go through the birthing pains. He also reviewed the chapters on signal representations and discrete-time filters. Greg Bashford, who should have been in my department but for some unknown reason preferred to spend his time in Biosystems Engineering at the University of Nebraska, reviewed the chapters on Computed Tomography and Magnetic Resonance Imaging. Kevin Pitt at the University of Nebraska looked over the chapter on EEG, focusing on the material related to Brain-Computer Interfaces. Sinan Sayood of Stanford University reviewed the chapter on the electrocardiogram. Amirsalar Mansouri of Louisiana State University tried to make the chapter on EEG more informative. And Sena Sayood of Washington University in St. Louis reviewed the second half of the book, identifying errors of commission and omission. My thanks to all of them for their careful review. Any remaining errors are entirely due to my inability to follow recommendations.

My thanks also to Joel Claypool, who encouraged me to write this book, and to Charles Glaser for graciously accommodating my delays. When I told him I could submit the manuscript on deadline, but it would not be something very satisfactory, his response was "I think we should make quality paramount." I hope this book will be closer to his standards than the on-time version would have been. Finally, and always, I would like to thank my life partner Fusun, without whom I would not be a functioning human being.

Lincoln, USA Khalid Sayood

Competing Interests The author has no competing interests to declare that are relevant to the content of this manuscript.

Contents

Introduction

1

Signal processing involves extracting information from signals. Signals are everywhere, from the movement of leaves on a tree to the smile on a person's face; anything that conveys information about our surroundings, our person, or our reality is a signal. Signal processing deals with more prosaic signals, those that can be measured and recorded. Of these signals, those related to our health and bodies are classified as biomedical signals. Even with these restrictions, the number of different kinds of biomedical signals is enormous. A complex system cannot operate without the ability to transfer information between its component systems. The operation of the complex system itself will generate various signals. The human body is a highly complex system composed of numerous intricate subsystems. For such a complex system to function, let alone function efficiently, an enormous number of signals conveying a massive amount of information must be constantly generated, evaluated, and acted upon. Therefore, we do not lack biomedical signals on which to practice our signal processing techniques. However, we are limited in this book to what we can cover in one semester; hence, our focus is on the most widely used signals.

We will focus on examples of two types of signals: those generated by the body and those generated during the imaging process of our body. Examples of the first kind of signals include the electroencephalograph (EEG) and the electrocardiogram (EKG). Examples of the second kind of signal include those generated by computed tomography (CT) and magnetic resonance imaging (MRI). These are hardly the only signals, or even the only types of signals, that fall under the category of biomedical signals. Other biomedical signals include blood pressure variations, pulse oximetry, signals from muscles or electromyograms (EMG), signals from peripheral nerves or electroneurograms (ENG), gene expression signals from transcriptional profiling, electro-oculograms (EOG), galvanic skin resistance, ultrasound signals generated for several imaging applications, and many, many more. Hope-

K. Sayood, *Biomedical Signal Processing*, Synthesis Lectures on Biomedical Engineering,
https://doi.org/10.1007/978-3-032-08855-0_1

fully, studying the signal processing approaches used in the limited set of examples in this book will provide us with tools that we can also apply to other biomedical signals.

The more "standard" signals that we are used to in signal processing, such as speech, images, and video, all have characteristics that are dictated by the systems that generate them and selected for by the systems that receive them. Speech signals possess distinct characteristics due to the structure and limitations of the speech production apparatus. The vibrations of the vocal folds introduce a periodicity or pitch into the speech signal. The limitations on the flexibility of the path traveled by the speech signal do not permit rapid changes and hence limit the range of frequencies. While audio signals, such as music, are not limited by the vocal apparatus, they are still subject to selective limitations imposed by human hearing preferences and limitations. The range of frequencies that humans can hear is limited, so it makes sense to use instruments that generate sounds within that range. With images, pictures selected for viewing by people are those that are suitable for the band-limited characteristic of the human visual system. Humans tend to focus more on edges in pictures, so image processing approaches will penalize distortion in edge pixels more heavily than in other regions of the image. The fact that not all frequencies are visible allows compression algorithms, such as JPEG and MPEG, to discard high-frequency information that is not perceptible.

Similarly, the biomedical signals we deal with are also shaped by how they are generated and the environment in which they are utilized. The signal processing of biomedical signals also has a distinct focus from that of audio and video signals. Unlike audio and video signal processing, the goal of biomedical signal processing is not to make the signals pleasant to perceive. Instead, the goal is to extract information. Furthermore, unlike audio and video signal processing, which cater to a general audience, the audience for biomedical signal processing is narrower and more specialized. Ultimately, the various signals encountered in biomedical signal processing are highly diverse, and it is essential that we appreciate this diversity and design our algorithms accordingly.

Consider the EEG signals shown in Fig. 1.1. EEG signals initially appear to be noise, and there are stories, likely apocryphal, of experts being fooled into thinking that randomly generated noise was an EEG signal. However, we know that wherever there is information, there is structure. Extracting that structure and the information it contains is our job. In contrast, consider the EKG signal shown in Fig. 1.2. The structure is evident here. Our task here is to focus on detecting subtle variations in this structure and correlating them with various types of pathologies.

To accomplish our tasks, we require a wide variety and types of information. We will need an understanding of the physiological systems that generate the signals of interest, and we will require a variety of tools to probe and analyze these signals. The more tools we have, the more our ability to devise algorithms to accomplish the tasks we are interested in. Finally, we require an understanding of the technologies used to extract the signals of interest.

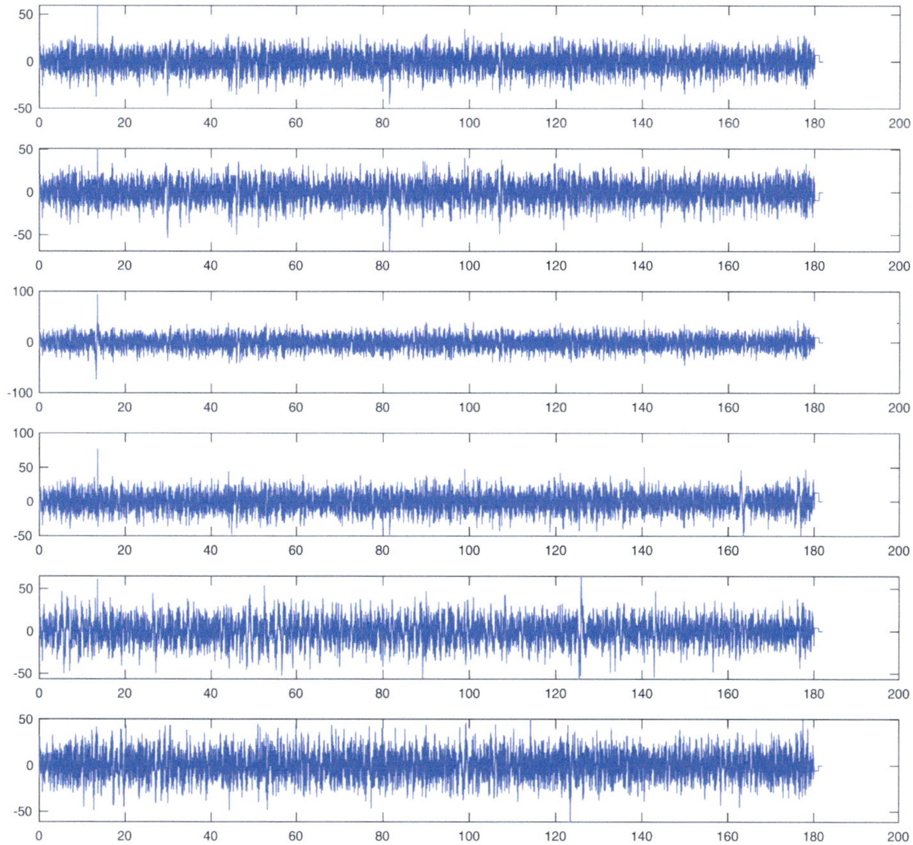

Fig. 1.1 EEG signals

Fig. 1.2 An EKG signal

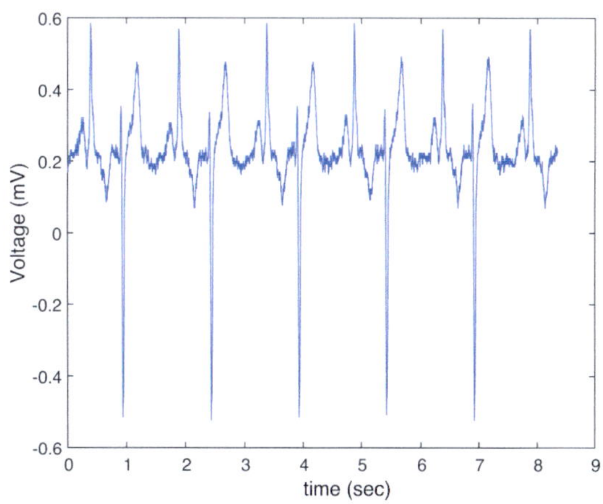

1.1 Organization of This Book

To cover these diverse issues, this book is divided into three parts. In the first part, consisting of Chaps. 2–5, we introduce some of the computational techniques that will be useful when examining the various signals of interest. The second part, comprising Chaps. 6–8, addresses physiological signals. In the third part, Chaps. 9 and 10, we examine two imaging modalities widely used to extract information about physiological function: computed tomography (CT) and magnetic resonance imaging (MRI). We briefly describe these parts below.

1.1.1 Computational Approaches

In Chap. 2, we examine signal representation in both the time and frequency domains. We assume that the reader has some familiarity with Fourier transforms. For those who need a quick refresher, we have included an appendix on the subject. The biomedical signals we are interested in—the EKG and the EEG—are analog, but almost all processing nowadays uses digital processors, so these analog signals need to be converted into digital signals. The first step in this conversion process is sampling, which involves transforming a continuous-time signal into a discrete-time signal. We examine some of the potential constraints and trade-offs involved in this process. After the samples have been obtained, they are quantized so their values can be represented using a fixed number of bits. Unlike the sampling process, the quantization process is irreversible; the information discarded during the quantization process cannot be recovered. Therefore, it is essential that we are aware of the distortion this process can introduce. We then look at discretization in the frequency domain. In particular, we examine the discrete Fourier transform (DFT) and the various issues associated with its use, as well as its most popular implementation, the Fast Fourier Transform (FFT). We then turn our attention to discrete-time filters in Chap. 3, which are often an essential part of biomedical signal processing. We briefly review the Z-transform and then examine some approaches for designing various types of discrete-time filters. With the tools available nowadays, discrete-time filters are straightforward to design and implement. Knowing some of the underlying theory can make one feel more comfortable working with these filters.

A significant portion of biological signal processing today involves classification, sometimes under the label of *machine learning*. Classification schemes broadly consist of two types of activities: the extraction of features that allow us to differentiate between classes and the partitioning of the feature space into classes. Features can be specific to particular signals and applications, such as heart rate for cardiac signals, or they can be application-agnostic. We examine various methods for extracting application-agnostic features from signals in Chap. 4. The types of features we examine include statistical features, such as mean, variance, and correlation, as well as model-based features. Additionally, we consider matrix-based features, which can be obtained using Singular Value Decomposition (SVD) and Principal Component Analysis (PCA). In Chap. 5, we look at how these features can be utilized in classification. We examine two popular classification algorithms: the Generalized Lloyd algorithm, a close cousin of the k-means algorithm, and Support Vector Machines.

1.1.2 Physiological Signals

The two physiological signals we examine in some detail, in Chaps. 7 and 8, are the electrocardiogram and the electroencephalogram. Both of these signals are based on the transmission of signals through neurons. It is essential to understand that while the EKG and the EEG are electrical signals, the signals that regulate the beating of the heart and the activities in the brain are electrochemical. They do not behave in the way signals we are more used to, such as signals traveling over wires, or cables, or through the atmosphere. In Chap. 6, we briefly examine how these neuronal signals are generated. In Chap. 7, we provide a brief introduction to the physiological basis of ECG generation, the practicalities of measuring the signal, and the terminology used to describe it. We then examine various signal processing approaches for artifact removal and feature extraction, which can be utilized to aid in the identification of different pathologies. In Chap. 8, we follow a similar approach to studying the EEG signal. We briefly review the physiological basis for generating the EEG signal and standard approaches to measuring it. The frequency profile of the EEG signal is of particular interest, as different brain states are reflected in distinct frequency profiles. Similar to EKG, we examine signal processing approaches for artifact removal and then consider the use of EEG in detecting brain states, both normal and pathological.

1.1.3 Imaging Modalities

In Chaps. 9 and 10, we transition from examining signals generated by the body to signal processing approaches used to study the body. We describe two popular imaging modalities, Computed Tomography and Magnetic Resonance Imaging. In each case, we examine how the images are generated, with a primary focus on the signal processing involved in creating the images. While both are imaging modalities with the same target—the human body—and both are highly innovative, the approaches for generating the images are quite different. In CT, images are reconstructed from projections; in MRI, images result from nuclear magnetic resonance, with magnetic gradients used to distinguish signals from different physical locations. Each utilizes distinct aspects of signal processing, demonstrating both the diversity and capabilities of this field.

The field of biomedical signal processing combines fascinating aspects from multiple areas of study. There is the complexity of the signals, which reflect the complex nature of the underlying systems. Signal processing enables us to identify the relationships between the various features of the signal and the underlying biological process. To do this, one needs to have some appreciation for the biological processes that are generating the signals, as well as the tools that allow us to study these processes. Biomedical signal processing is the ideal playground for those interested in the capabilities of signal processing algorithms and the processes that sustain life.

2 Signal Representation

According to one of the many definitions in the Cambridge dictionary, a signal is "something that shows something else exists or is likely to happen." For us that first "something" is generally a change in a measurable quantity such as a voltage, a current, or a pressure, amongst other things, which carries information about the second "something" which may be the heart, or the lung, or the brain. Within the human body, these signals are often those that help us stay alive, such as the signals that regulate the beating of the heart, the inhalation and exhalation of breath from the lungs, or the firing of neurons in the brain. On the outside, the signals we are interested in help us investigate the functions of the body, such as the signals that indicate the presence of ionizing radiation, like X-rays, or the signal that represents the magnetic resonance of different molecules, or the signal that represents the strength of reflected ultrasound waves. All these signals need a representation that can be easily accessed, stored, and processed. Generally, these representations are functions of time and/or space. Once we have a temporal or spatiotemporal representation of the signal, we can find other representations, such as frequency domain representations using transform techniques. We will examine these signals in more detail in later chapters. In this chapter, our interest is in the representation of these signals.

Until recently (depending on the definition of recently), most of our machinery for processing signals was analog; therefore, our representations of signals were also analog. Now, almost all our processing is done using digital discrete-time processors, so we need to represent our signals digitally as well. Let's examine the transition from the world of analog signals to the world of digital signals.

K. Sayood, *Biomedical Signal Processing*, Synthesis Lectures on Biomedical Engineering,
https://doi.org/10.1007/978-3-032-08855-0_2

2.1 From Analog to Digital

For the moment, let's focus on signals that are functions of time. A time-domain representation of the signal $x(t)$ is analog in two distinct ways. The time variable t is continuous and can take on values over a continuum. The same is true of the amplitude of the signal $x(t)$. While the amplitude may be bounded—and it usually is—within the bound, it can take on an infinite number of values. To transition from analog to digital, we must discretize both time and amplitude. We can view this conceptually as a two-step process; first, we discretize the time, and then we discretize the amplitude. We discretize the time through the *sampling* process, and we discretize the amplitude through the *quantization* process.

2.1.1 Sampling

The process of discretizing the signal in time is called *sampling*. From the signal $x(t)$, we want to obtain the signal $x(nT_s)$, where T_s is the interval between samples and n now represents our time index. It seems that when we sample a signal, we ultimately have to discard information. After all, keeping only samples of a signal means that all the points we did not sample are discarded. So, exact recovery of the original signal from the samples should be impossible. Fortunately, as we will see below, this is not true, and we can recover the original signal exactly if we pick our sampling interval T_s appropriately. It should be clear that if the signal $x(t)$ is changing rapidly, we need to sample it more often to ensure that we capture the changes, rather than if the signal were changing slowly. The extreme case would be a constant signal for which a single sample will suffice. A measure of the rate of change of the signal is the *bandwidth* of the signal. As we shall see, we need to select a sampling rate that is directly proportional to the bandwidth; the higher the bandwidth (and hence the rate of change), the higher the required sampling rate. As the interval between samples gets smaller the faster we sample the signals, we can also say that the sampling interval T_s is inversely proportional to the bandwidth.

To construct a mathematical model for the sampling process, we need a function that is non-zero only at discrete points in time. An impulse function $\delta(t)$ is a function that is non-zero only for $t = 0$. We can put together shifted versions of the impulse function into an impulse train $\sum \delta(t - kT_s)$, where T_s is the interval between the impulses. The sampled signal $x_s(t)$ can then be modeled as a product of the original signal $x(t)$ with the impulse train.

$$x_s(t) = x(t) \cdot \sum_{k=-\infty}^{\infty} \delta(t - kT_s)$$

A visualization of this process is shown in Fig. 2.1.

To see what the effect of this (ideal) sampling is, we need to look at the process in the frequency domain. We will utilize the Fourier transform, specifically its multiplication

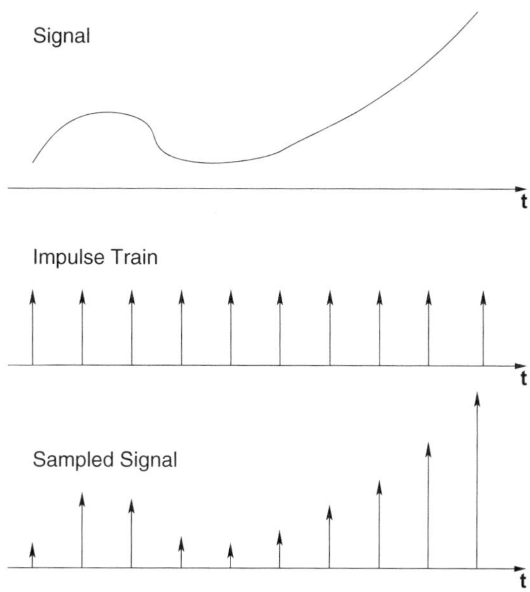

Fig. 2.1 Representing sampling as the product of the signal with an impulse train

property, to examine what occurs during the sampling process. (If you are not familiar with the Fourier transform, take a look at Appendix A for a brief overview.) According to the multiplication property, the Fourier transform of the product of two signals is the convolution of their individual Fourier transforms.

$$\mathcal{F}\left[a(t) \cdot b(t)\right] = \mathcal{F}\left[a(t)\right] \circledast \mathcal{F}\left[b(t)\right]$$

As the sampled signal is the product of the original analog signal $x(t)$ and the impulse train, the Fourier transform of the sampled signal will be the convolution of $X(\omega)$, the Fourier transform of $x(t)$, and the Fourier transform of the impulse train.

$$X_s(\omega) = \frac{1}{2\pi} X(\omega) \circledast \mathcal{F}\left[\sum_{k=-\infty}^{\infty} \delta(t - kT_s)\right]$$

The Fourier transform of an impulse train is an impulse train.

$$\mathcal{F}\left[\sum_{k=-\infty}^{\infty} \delta(t - kT_s)\right] = \frac{2\pi}{T_s} \sum_{k=-\infty}^{\infty} \delta(\omega - k\omega_s)$$

where

$$\omega_s = \frac{2\pi}{T_s}$$

so

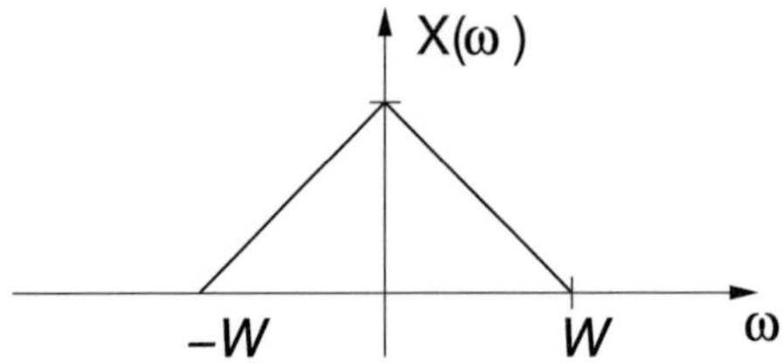

Fig. 2.2 The Fourier transform of the original signal

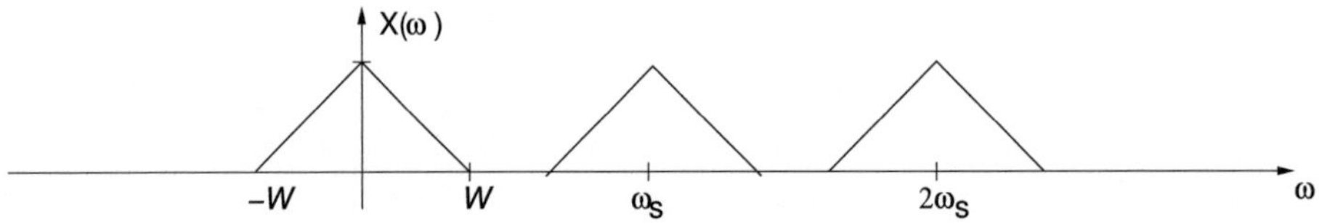

Fig. 2.3 Three terms in the summation

$$X_s(\omega) = \frac{1}{2\pi} X(\omega) \circledast \frac{2\pi}{T_s} \sum_{k=-\infty}^{\infty} \delta(\omega - k\omega_s)$$

or

$$\begin{aligned} X_s(\omega) &= \frac{1}{T_s} \sum_{k=-\infty}^{\infty} X(\omega) \circledast \delta(\omega - k\omega_s) \\ &= \frac{1}{T_s} \sum_{k=-\infty}^{\infty} X(\omega - k\omega_s) \end{aligned}$$

What does this look like? Let's assume the shape shown in Fig. 2.2 for $X(\omega)$. Let's now plot a few of the terms from the summation; say $X(\omega) + X(\omega - \omega_s) + X(\omega - 2\omega_s)$. The term $X(\omega - \omega_s)$ is simply $X(\omega)$ shifted to ω_s and similarly $X(\omega - 2\omega_s)$ is $X(\omega)$ shifted to $2\omega_s$. When we add these shifted replicas together, we obtain the frequency domain representation shown in Fig. 2.3.

The Fourier transform of the sampled signal $X_s(\omega)$ shown in Fig. 2.4 is the sum of all the shifted replicas of the original transform $X(\omega)$ where the shifts are integer multiples of ω_s.

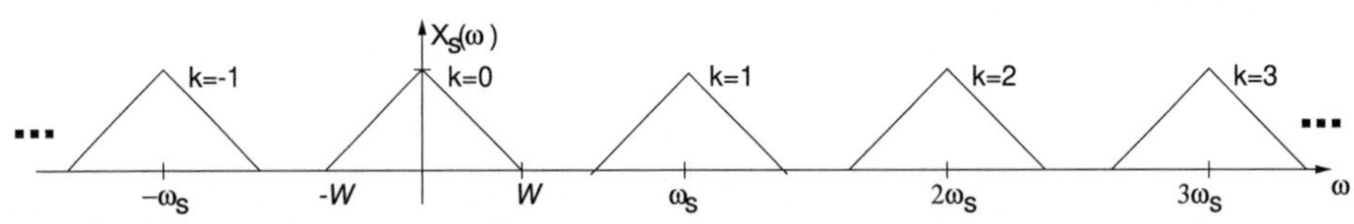

Fig. 2.4 Fourier transform of the sampled version of the signal shown in Fig. 2.2

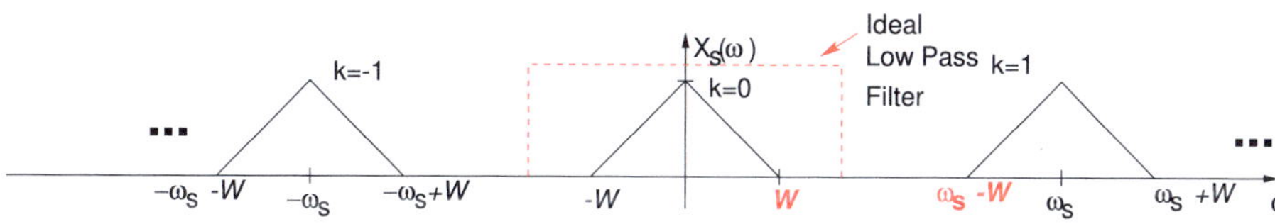

Fig. 2.5 An expanded view of three replications of the Fourier transform of the sampled signal

Looking at Fig. 2.4, we can see how we can recover our original signal from this sampled signal. We need to eliminate all the replicas. We can do that easily by using a low-pass filter with a bandwidth greater than W and less than $\omega_s - W$.

So, where is the catch? The reason we could recover the original signal using a low-pass filter was that the replicas were all separated in frequency. In Fig. 2.5, we have plotted an expanded view of three replications. From Fig. 2.5 we can see that the reason the replicas are separated in frequency is that the sampling frequency ω_s was such that the lower end of the first replica at $\omega_s - W$ was located at a frequency higher than W, or

$$\omega_s - W > W$$

or $\omega_s > 2W$. This is a requirement for the sampling frequency if we want to recover our original signal. The component at W is the highest frequency in the signal, so one way to state this requirement is to say that the sampling rate has to be greater than twice the highest frequency contained in the signal. This frequency is called the *Nyquist frequency* after Harry Nyquist, who first discovered it [1, 2]. In practice, we would want to sample at a higher rate, as the larger the value of ω_s, the further apart the replicas will be. The further apart the replicas are, the more relaxed the requirements on the reconstruction filter.

The signals we have drawn to illustrate our development have been low-pass signals—signals that have components at frequencies that range from zero to the bandwidth of the signal. But what if the original signal is not a low-pass signal, but a band-pass signal? A bandpass signal is a signal with frequency components that occupy a band from ω_l to ω_h where ω_l is greater than zero. The bandwidth W of the bandpass signal is the range of frequencies it occupies, $\omega_h - \omega_l$. An example of the frequency profile of a bandpass signal is shown in Fig. 2.6.

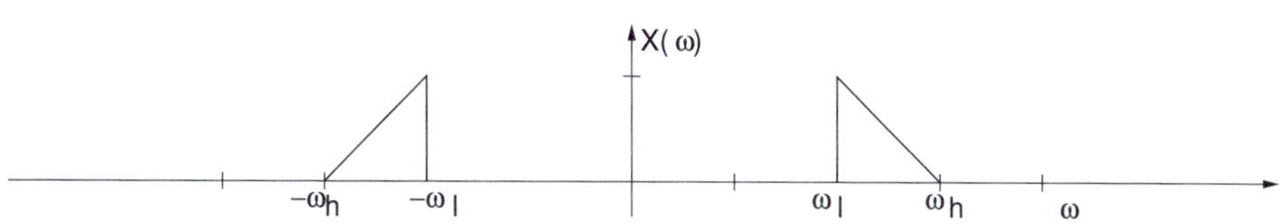

Fig. 2.6 A bandpass signal with non-zero frequency components between ω_l and ω_h

The highest frequency component in this signal is at ω_h. If we sampled this signal at a rate greater than twice the highest frequency, we would recover the original signal. However, in this case, we have space in the frequency domain between zero and ω_l so, as long as the bandwidth W is smaller than ω_l, we could sample this signal in such a way that the replicas do not overlap the original signal without having to sample at rates greater than twice the highest frequency. Consider the example shown in Fig. 2.7 for a case where the sampling frequency is less than twice the highest frequency ω_h.

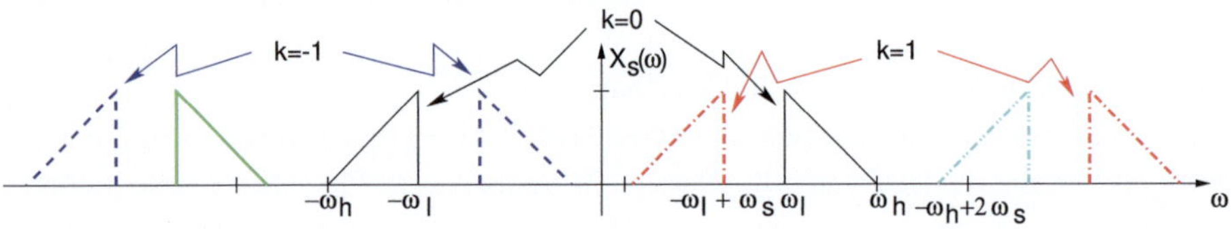

Fig. 2.7 A sampled bandpass signal with sampling frequency ω_s less than $2\omega_h$. We have drawn the different replicas with different line styles and colors

So, as long as $W \leq \omega_l$, what should be the requirements on the sampling frequency? Let's use the example in the figure to determine the requirement for the sampling frequency ω_s. From the figure, it is evident that if we do not want the replica to impinge on the frequencies contained in the original signal, for there to be no overlap between the original signal and the replica, we need two conditions:

$$-\omega_l + \omega_s < \omega_l \Rightarrow \omega_s < 2\omega_l$$
$$-\omega_h + 2\omega_s > \omega_h \Rightarrow 2\omega_s > 2\omega_h$$

Multiplying the top equation by –1, we get

$$-\omega_s > -2\omega_l$$
$$2\omega_s > 2\omega_h$$

Notice that multiplication by –1 also reversed the inequality in the first equation. Now, adding the two equations, we obtain the requirement on the sampling frequency

$$\omega_s > 2(\omega_h - \omega_l)$$

The sampling frequency has to be greater than twice the bandwidth of the signal.

We determined the sampling frequency requirement by examining the specific case in the figure, where the replica below the original signal was the first replica. What about the general case where the replica below the original signal is the kth replica? In this case, the replica below ω_l has to satisfy

$$-\omega_l + k\omega_s \leq \omega_l$$

where k is an integer. The replica just below the original signal will be the one for which k is maximum while satisfying this equation. We can find this value of k by noting that

$$k\omega_s \leq 2\omega_l \Rightarrow k \leq \frac{2\omega_l}{\omega_s}$$

As k is an integer, this means that the largest value of k satisfying this condition is given by

$$k = \left\lfloor \frac{2\omega_l}{\omega_s} \right\rfloor$$

The next replica of the original signal will be at $(k+1)\omega_s$. For this replica not to overlap with the original signal, we need

$$-\omega_h + (k+1)\omega_s \geq \omega_h$$

or

$$2\omega_h \leq (k+1)\omega_s$$

To find ω_s that satisfies both conditions

$$\begin{aligned} 2\omega_h &\leq (k+1)\omega_s \\ 2\omega_l &\leq k\omega_s \end{aligned}$$

subtract the second equation from the first, and we again get

$$2(\omega_h - \omega_l) \leq \omega_s$$

or the sampling frequency should be greater than twice the bandwidth of the original signal. Recovery of the signal in this case requires the use of bandpass filters, so we need to know both ω_l and ω_h.

It turns out that sampling does not necessarily result in a loss of information, as long as the sampling rate is at least twice the bandwidth of the signal. The distortion caused by sampling at rates lower than the required rate is called *aliasing*, and in most real-world situations, eliminating it is impossible. There are times when we are stuck with a particular sampling frequency that is less than twice the bandwidth of the signal being sampled due to the processor we must use, or other constraints. This is the situation when sampling voice signals for transmission over telephones. Due to various historical reasons, the most commonly used sampling rate for voice communication is 8,000 samples per second. This would be fine if all frequency components in the voice signal were at frequencies below 4 kHz. However, the voice contains components at least up to 8 kHz. Sampling this signal at 8,000 samples per second would result in a considerable amount of aliasing. Fortunately for us, most of the information required for intelligibility in the voice signal is contained in frequencies below 4 kHz. Therefore, to avoid the aliasing distortion, we first filter the voice signal with an *antialiasing filter* with a cutoff at 3.6 kHz. Note that this still introduces distortion; the components above 4 kHz significantly contribute to the audio quality of the voice signal. However, this distortion is considerably less than the aliasing distortion. Additionally, it involves controlled distortion. The use of the antialiasing filter is intended to minimize our losses. Some voice-over-Internet protocols (VoIP) utilize a higher sampling rate, which can provide a higher-quality voice. Where quality is essential, for example, for music signals, we use a much higher sampling rate, typically 44,100 samples per second.

Why not use a higher sampling rate consistently? Firstly, an increased sampling rate requires increased processor power, which may not necessarily be available or convenient. Secondly, each sample must be quantized and represented by several bits, as we will see in the next section. Increasing the sampling rate results in an increase in the bit rate, which in turn means a higher bandwidth for transmission. Ultimately, it is a matter of tradeoffs.

2.1.2 Quantization

Sampling discretizes the signal in time—an essential step if we are to use digital processing. The samples also need to be discretized in amplitude because a specific number of bits will represent each sample, and a fixed number of bits can only represent a limited number of values. In practice, the samples are represented by digital codewords using an Analog-to-Digital (A/D) converter. These digital codewords can be transformed back to analog values using a Digital-to-Analog (D/A) converter. The concatenation of these processes is called *quantization* (Fig. 2.8). The A/D divides the range of values assumed to be taken on by the sample into subintervals. The number of subintervals depends on the number of bits per sample. If we are to use eight bits per sample, as is generally the case with cell phones, the number of subintervals is 2^8, or 256. As an example, in Fig. 2.9 we show the case where we are using three bits per sample.

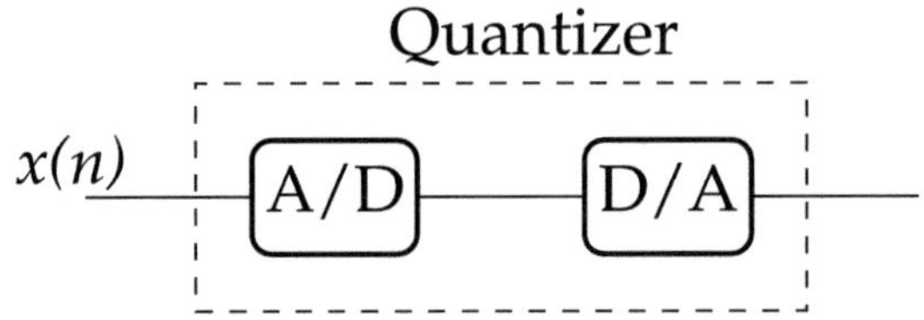

Fig. 2.8 The quantizer consists of the concatenation of an A/D and a D/A

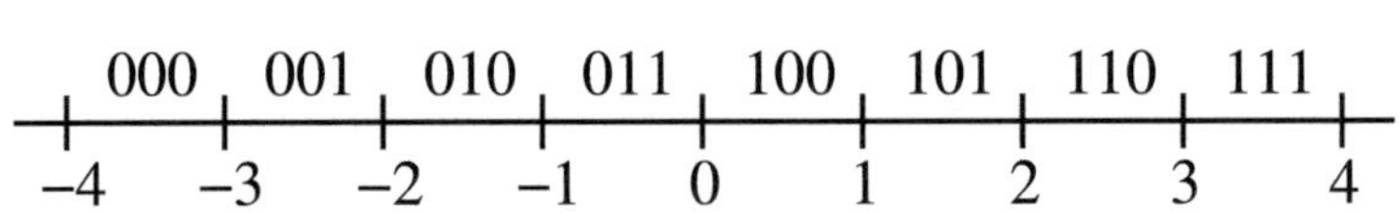

Fig. 2.9 A three bit A/D

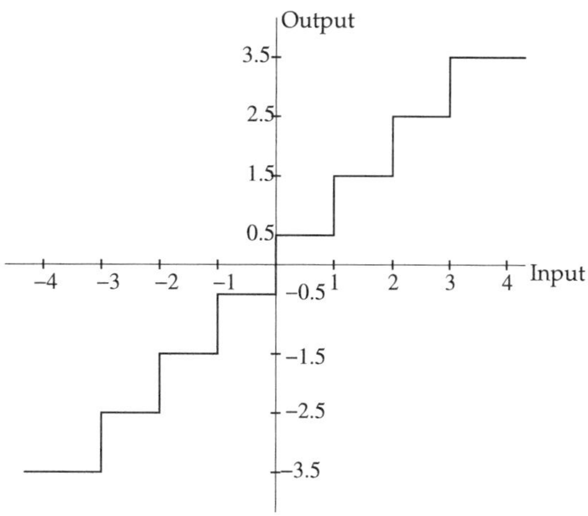

Fig. 2.10 A three-bit quantizer

In this example, if the sample had a value between –4 and –3, the A/D would put out the code 000. If the sample had a value between 1 and 2, the A/D would put out a code 101. Notice that the code represents the interval in which the sample value resides, not the precise value of the sample. Therefore, a sample value of 1.1 would result in a code of 101, as would a sample value of 1.65. The D/A takes the code and generates a value. As the code only denotes the interval, the output of the D/A must represent all values within that interval. Generally, the midpoint of the interval is taken as the representative value. Therefore, given code 101, the D/A would output a value of 1.5. This means that if the sample value were 1.1, the reconstruction value would be 1.5, and if the sample value were 1.65, the reconstruction value would still be 1.5. The input/output relationship of this particular concatenation of A/D and D/A, the quantizer is given in Fig. 2.10. Notice that the intervals in this case are all the same size. This reflects an assumption that the sample value is equally likely to be any value between –4 and 4. Quantizers of this kind are referred to as uniform quantizers.

Denoting the sample value, which is the input to the quantizer, by $x(n)$, and the reconstruction value, which is the output of the quantizer, by $y(n)$, we can define the quantization error as

$$q(n) = x(n) - y(n)$$

For this case, the quantization error will vary between –0.5 and 0.5 when the input falls between –4 and 4, and will grow without bound outside this region as shown in Fig. 2.11.

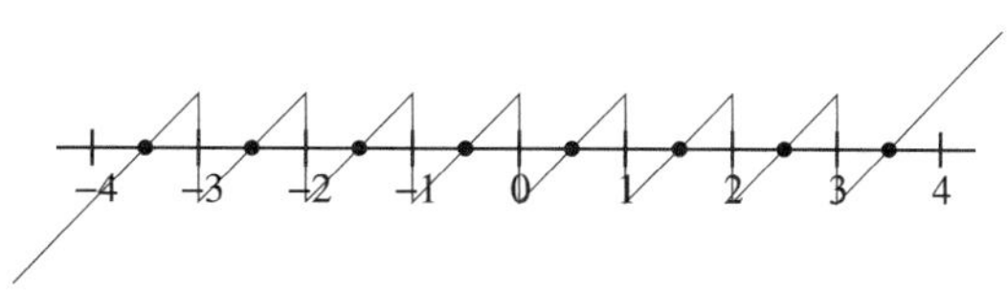

Fig. 2.11 The quantization error for a three-bit uniform quantizer

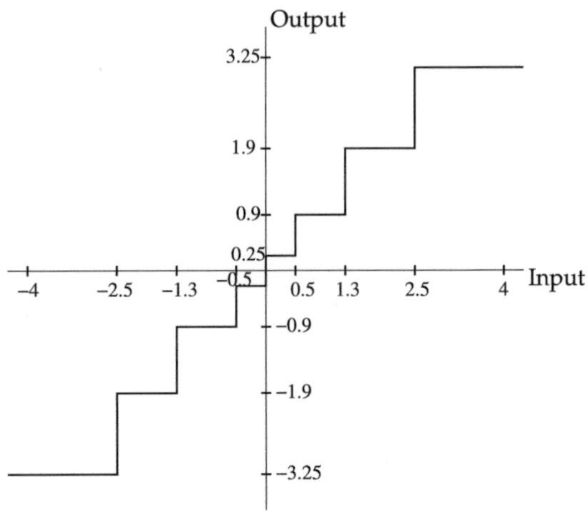

Fig. 2.12 A three-bit non-uniform quantizer

Often, the sample values are not uniformly distributed over the entire range of the signal. When dealing with speech, for example, smaller magnitude values of the sample are more likely than larger magnitude values. In this case, it makes sense to use a nonuniform quantizer in which the quantization intervals near the origin are smaller than those further from the origin. A three-bit example is shown in Fig. 2.12. In this case, smaller sample values are reconstructed with smaller quantization errors than larger sample values. For example, in this case, if the sample has a value between 0 and 0.5, the reconstruction value is 0.25, and the quantization error is between –0.25 and 0.25. If the sample has a value between 2.5 and 4, the reconstruction value is 3.25, and the quantization error lies between –0.75 and 0.75. Notice that in the latter case, the range of the quantization error is larger than it was for the uniform quantizer. However, if our assumption about the distribution of the sample values is correct, more of the samples will fall into the inner quantization levels than into the outer quantization intervals. Therefore, most of the quantization errors will be smaller for the non-uniform quantizer than for the uniform quantizer.

2.2 Discretization in Frequency

Up until now, what we have done is digitize the time signal. However, when we look at the Fourier transform of the sampled signal, we can see that it is a continuous function of the frequency ω. To use digital signal processing, the frequency representation of the signal should also be digital. We look at the discretization of the frequency representation in two steps. First, we find the Fourier transform of the sampled discrete-time sequence. This is called the discrete-time Fourier transform (DTFT). We then sample the DTFT to find the discrete Fourier transform (DFT).

As we saw above, the sampled discrete time sequence can be modeled as the product of the original time function $x(t)$ and an impulse train $\sum_{n=-\infty}^{\infty} \delta(t - nT)$. Therefore, the Fourier transform of the discrete-time function can be written as

$$\mathcal{F}\left[x(t) \sum_{n=-\infty}^{\infty} \delta(t - nT)\right] = \int_{-\infty}^{\infty} x(t) \sum_{n=-\infty}^{\infty} \delta(t - nT)e^{-j\omega t}dt$$
$$= \sum_{n=-\infty}^{\infty} \int_{-\infty}^{\infty} x(t)\delta(t - nT)e^{-j\omega t}dt$$
$$= \sum_{n=-\infty}^{\infty} x(nT)e^{-j\omega nT}$$

where we have used the sifting property of the delta function,

$$\int_{-\infty}^{\infty} x(t)\delta(t - t_o)dt = x(t_o)$$

If we normalize the sampling interval to one, the DTFT $X(e^{j\omega})$ can be written as

$$X(e^{j\omega}) = \sum_{n=-\infty}^{\infty} x[n]e^{-jn\omega}$$

where we have replaced parentheses with brackets to emphasize that we are now dealing with a discrete sequence. Recall from above that the Fourier transform of the discrete-time function is periodic with period $\omega_S = 2\pi/T$. We have set $T_s = 1$; therefore, the DTFT will be periodic with period 2π.

To obtain the inverse transform, we use the following relationship

$$\frac{1}{2\pi}\int_0^{2\pi} e^{-jn\omega}e^{jk\omega}d\omega = \begin{cases} 1 & n = k \\ 0 & n \neq k \end{cases}$$

If we multiply both sides of the equation for $X(e^{j\omega})$ shown above with $\frac{1}{2\pi}e^{jk\omega}$ and integrate with respect to ω we obtain

$$\frac{1}{2\pi}\int_0^{2\pi} X(e^{j\omega})e^{jk\omega}d\omega = \frac{1}{2\pi}\int_0^{2\pi} \sum_{n=-\infty}^{\infty} x[n]e^{-jn\omega}e^{jk\omega}d\omega$$

Switching the order of integration and summation on the right-hand side, we obtain

$$\frac{1}{2\pi}\int_0^{2\pi} X(e^{j\omega})e^{jk\omega}d\omega = \sum_{n=-\infty}^{\infty} \frac{1}{2\pi}\int_0^{2\pi} x[n]e^{-jn\omega}e^{jk\omega}d\omega$$

$x[n]$ is not a function of ω so we can take it out of the integral

$$\frac{1}{2\pi}\int_0^{2\pi} X(e^{j\omega})e^{jk\omega}d\omega = \sum_{n=-\infty}^{\infty} x[n]\frac{1}{2\pi}\int_0^{2\pi} e^{-jn\omega}e^{jk\omega}d\omega$$

The integral on the right is zero for all values of n except $n = k$; therefore,

$$\frac{1}{2\pi}\int_0^{2\pi} X(e^{j\omega})e^{jk\omega}d\omega = x[k]$$

The discrete-time Fourier transform pair is

$$X(e^{j\omega}) = \sum_{n=-\infty}^{\infty} x[n]e^{-jn\omega}$$

$$x[n] = \frac{1}{2\pi}\int_{2\pi} X(e^{j\omega})e^{jn\omega}d\omega.$$

2.2.1 Discrete Fourier Transform

The reason we are interested in this discrete representation is that much of the processing nowadays is conducted in the discrete domain. This means that not only is the time domain representation of the signal discrete, but so are the frequency representations. The discrete-time Fourier transform is a transform for discrete-time functions—the representation in the time domain is discrete. However, the transform itself is a continuous periodic function of ω. To obtain a transform that provides discrete representations in both the time and frequency domains, we need to discretize the frequency to get the Discrete Fourier Transform (DFT).

The DTFT is given by

$$X(e^{j\omega}) = \sum_{n=-\infty}^{\infty} x[n]e^{-jn\omega}$$

The function $e^{j\omega}$ is periodic with period 2π, therefore, if we sample $X(e^{j\omega})$ N times over one period we will get

$$X(e^{jk(2\pi/N)}) = \sum_{n=-\infty}^{\infty} x[n]e^{-jnk(2\pi/N)}$$

As in the time domain, when we sample in the frequency domain, it appears that we are discarding a significant amount of information. Before sampling, we could obtain the value of $X(e^{j\omega})$ for an infinite number of values of ω between 0 and 2π. After sampling, we only have N of the infinite number of values available. When we sampled in time, we observed that we could recover the original infinite-valued signal if the Fourier transform satisfied the constraint that the bandwidth of the original signal was below half the sampling frequency. Are there constraints we can place on the time domain signal that will allow us to recover the signal $x[n]$ from the samples of the discrete-time Fourier transform (DTFT)? As you might expect, the answer is yes. We can recover $x[n]$ from $X(e^{jk(2\pi/N)})$ if $x[n]$ is nonzero for only L consecutive values of n where $L \leq N$.

Let's begin our justification of this statement by noting that

$$\sum_{n=0}^{N-1} e^{jk(2\pi/N)n} = \begin{cases} N & k = 0, \pm N, \pm 2N, \ldots \\ 0 & \text{otherwise} \end{cases}$$

We can see that this is true by using the geometric sum formula

$$\sum_{n=k}^{m} \alpha^k = \frac{\alpha^k - \alpha^{m+1}}{1-\alpha}$$

Letting α be $e^{jk(2\pi/N)}$ for $k \neq 0, \pm N, \ldots,$

$$\begin{aligned} \sum_{n=0}^{N-1} e^{jk(2\pi/N)n} &= \sum_{n=0}^{N-1} \left(e^{jk(2\pi/N)}\right)^n \\ &= \frac{e^0 - \left(e^{jk(2\pi/N)}\right)^{(N-1+1)}}{1 - \left(e^{jk(2\pi/N)}\right)} \\ &= \frac{1 - e^{jk2\pi}}{1 - \left(e^{jk(2\pi/N)}\right)} \\ &= 0 \end{aligned}$$

where we have used the fact that $e^{jk2\pi} = 1$.

What happens when $k = 0, \pm N, \ldots$? When $k = 0$

$$e^{jk(2\pi/N)} = e^0 = 1$$

When $k = mN$ for an integer value of m

$$e^{jmN(2\pi/N)n} = e^{j2\pi mn} = 1$$

Therefore, for these values of k

$$\sum_{n=0}^{N-1} e^{jk(2\pi/N)n} = \sum_{n=0}^{N-1} 1 = N$$

With this piece of information in our hands, let's look at what happens if $x[n] = 0$ for $n \geq N$. In this case, we can rewrite the limits of summation in our expression for $X(e^{jk(2\pi/N)})$ as

$$X(e^{jk(2\pi/N)}) = \sum_{n=0}^{N-1} x[n]e^{-jnk(2\pi/N)}$$

If we now multiply both sides of the equation with $e^{jlk(2\pi/N)}$ and sum from $k = 0$ to $N - 1$ we get:

$$\sum_{k=0}^{N-1} X(e^{jk(2\pi/N)})e^{jlk(2\pi/N)} = \sum_{k=0}^{N-1}\sum_{n=0}^{N-1} x[n]e^{-jk(2\pi/N)n}e^{jlk(2\pi/N)}$$

We can simplify the right-hand side by combining the two exponentials.

$$\begin{aligned}\sum_{k=0}^{N-1} X(e^{jk(2\pi/N)})e^{jlk(2\pi/N)} &= \sum_{k=0}^{N-1}\sum_{n=0}^{N-1} x[n]e^{j(l-n)(2\pi/N)k} \\ &= \sum_{n=0}^{N-1} x[n] \sum_{k=0}^{N-1} e^{j(l-n)(2\pi/N)k}\end{aligned}$$

The inner sum on the right-hand side is either equal to N or to 0, depending on the value of $(l - n)$. It is equal to N when $n - l = 0$ or $n = l$. For all other values of n, $(n - l) \neq 0, \pm N, \ldots$, therefore the inner sum is equal to 0 for all other values of n. Therefore,

$$\sum_{k=0}^{N-1} X(e^{jk(2\pi/N)})e^{jlk(2\pi/N)} = x[l]N$$

or

$$x[l] = \frac{1}{N}\sum_{k=0}^{N-1} X(e^{jk(2\pi/N)})e^{jlk(2\pi/N)}$$

We can recover the original time signal from samples of the frequency domain representation.

For convenience, let's define

$$X[k] = X(e^{jk(2\pi/N)})$$

Then we get the DFT equations. The equation that maps the sequence from the time domain to the frequency domain is given by

$$\boxed{X[k] = \sum_{n=0}^{N-1} x[n]e^{-jnk(2\pi/N)}}$$

and is called the forward DFT, and the equation that maps the frequency domain representation to the time domain is

$$\boxed{x[n] = \frac{1}{N}\sum_{k=0}^{N-1} X[k]e^{jnk(2\pi/N)}}$$

and is called the inverse DFT.

Therefore, given a discrete time sequence $\{x[n]\}$ which is zero for $n < 0$ and $n \geq N$, we can find the spectral representation at discrete frequency values. And given the discrete spectral representation $\{X[k]\}$, we can recover the discrete time sequence $\{x[n]\}$. But is this restriction on the values of $x[n]$ necessary? We can show the necessity of this restriction in several ways. An easy way to achieve this is to utilize the fact that, just as sampling in the time domain results in replication in the frequency domain, sampling in the frequency domain also results in replication in the time domain. To see this, let's use the values of $X[k], k = 0, 1, \ldots, N-1$ to obtain $x[n]$ for $n \geq N$.

$$x[N] = \frac{1}{N}\sum_{k=0}^{N-1} X[k]e^{jNk(2\pi/N)} = \sum_{k=0}^{N-1} X[k]e^{jk(2\pi)} = \sum_{k=0}^{N-1} X[k] = x[0]$$

$$x[N+1] = \frac{1}{N}\sum_{k=0}^{N-1} X[k]e^{j(N+1)k(2\pi/N)} = \frac{1}{N}\sum_{k=0}^{N-1} X[k]e^{jk(2\pi/N)} = x[1]$$

$$x[N+2] = \frac{1}{N}\sum_{k=0}^{N-1} X[k]e^{j(N+2)k(2\pi/N)} = \frac{1}{N}\sum_{k=0}^{N-1} X[k]e^{j2k(2\pi/N)} = x[2]$$

$$\vdots \quad \vdots$$

$$x[2N-1] = \frac{1}{N}\sum_{k=0}^{N-1} X[k]e^{j(2N-1)k(2\pi/N)} = \frac{1}{N}\sum_{k=0}^{N-1} X[k]e^{j(N-1)k(2\pi/N)} = x[N-1]$$

Continuing in this manner, we can demonstrate that the recovered sequence $\{x[n]\}$ is periodic with a period of N. Just as sampling in the time domain resulted in a periodic extension in the frequency domain, sampling in the frequency domain results in a periodic extension in the time domain. Just as increasing the number of samples in the time domain spreads the replicas in the frequency domain, increasing the number of samples in the frequency domain spreads out the replicas in the time domain. And just as an insufficient number of samples (undersampling) in the time domain results in aliasing in the frequency domain, an insufficient sampling in the frequency domain results in aliasing in the time domain. The bottom line, though, is that if we have a time-limited sampled signal, we can get a unique spectral profile of it in the discrete domain. This enables us to perform operations in the spectral domain using digital processors that we would otherwise be unable to do.

Another conclusion we can draw from the above is that implicit in the DFT is the assumption that the sequence $\{x[n]\}$ is periodic with period N. We will explore the implications of this in the next section.

In the DFT literature, an additional notational shortcut is often used. Define

$$W_N = e^{-j\frac{2\pi}{N}}$$

then The forward DFT becomes

$$\boxed{X[k] = \sum_{n=0}^{N-1} x[n]W_N^{nk}}$$

and the inverse DFT becomes

$$\boxed{x[n] = \frac{1}{N}\sum_{k=0}^{N-1} X[k]W_N^{-nk}}$$

2.2.2 The Fast Fourier Transform

In practice, directly computing the DFT can be very resource-intensive, with the number of computations growing with N^2. Due to this, several fast algorithms have been developed, with the most popular being the Fast Fourier Transform (FFT) algorithm, developed by Cooley et al. [3]. (Unknown to them, Gauss had also come up with this algorithm almost 150 years before them—even before Fourier's work [4]). A description of the algorithm is beyond the scope of this course, but if you are interested in using it, MATLAB has an *fft* function you can play with. Most other widely used programming languages also have FFT routines you can use.

2.3 The Assumption of Periodicity

The discretization of the frequency imposes a periodicity condition on the input where the samples under consideration correspond to one period. Suppose the input is indeed periodic and the N samples make up an integer number of periods. In that case, this imposed periodicity coincides with reality, and the DFT agrees with what we would expect from the Fourier Transform. However, if this is not the case, we end up with some difficulties. To explore these issues and examine the use of the FFT, let us consider the FFT of a simple sinusoid. We will use the MATLAB routine *fft* for this investigation.

We know that the Fourier transform of $x(t) = \cos(2\pi f_o t)$ is given by $\pi\,[\delta(f - f_o) + \delta(f + f_o)]$. We begin by generating the discrete signal:

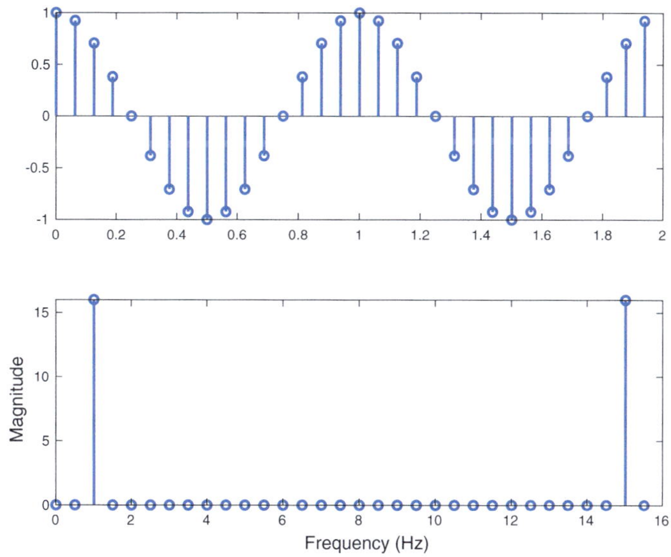

Fig. 2.13 FFT of $\cos(2\pi t)$ sampled at 16 samples per second using 32 samples

```
clear all
Ts = 1/16; % sampling interval
N=32; % Number of data points
fo = 1; % Frequency of the sinusoid
t = (0:1:N-1)*Ts; % sample points
x = cos(2*pi*fo*t); %Sampled sinusoid
```

When we sampled the DTFT, the distance between the samples in frequency was $2\pi/N$. However, we were using a value of $T_s = 1$. For a general value of the sampling interval T_s, the distance between samples would be $2\pi/NT_s$. In terms of cycles per second or Hz, the frequency interval, Δf, is $1/NT_s$. For this particular example $N = 32$ and $T_s = 1/16$, therefore, $\Delta f = 0.5$.

```
y = fft(x); % Take the FFT
f=(0:1:N-1)/(N*Ts); % Set up the frequency axis
```

The signal and the magnitude of the frequency coefficients are shown in Fig. 2.13.

The frequency profile shows the frequency components from 0 Hz to 15 Hz at intervals of 0.5 Hz. As expected, we have a nonzero coefficient at $f = 1$Hz. However, we also have a nonzero coefficient at $f = 15$Hz. Also, the nonzero coefficients both have a value of 16. To understand where the nonzero coefficient at 15 Hz comes from, recall that when we sampled the time signal with a sampling frequency of 16 samples per second, in the frequency domain, we got a replica of the original transform at every 16 Hz, as shown in Fig. 2.14. The Fourier

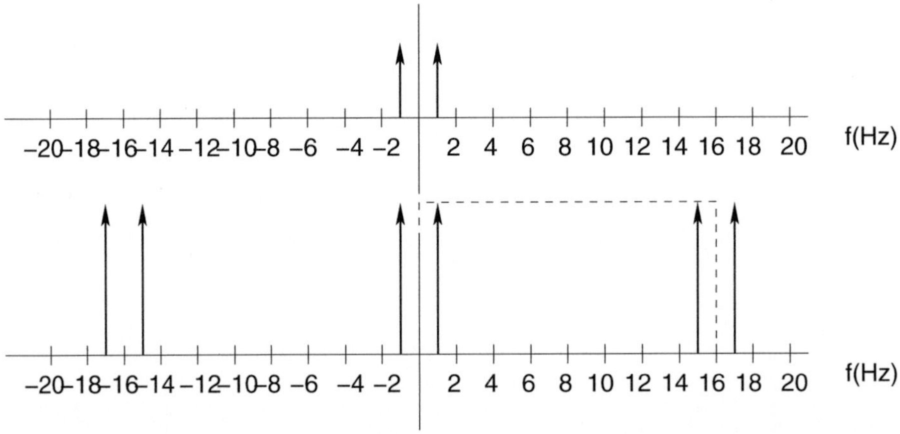

Fig. 2.14 Fourier transform a 1 Hz sinusoid and its sampled representation

transform of the original signal is delta functions at ±1 Hz. If we replicate this every 16 Hz, we will also get delta functions at 15 Hz, 17 Hz, 31 Hz, etc.

As we are only displaying the frequencies from 0 to 16 Hz, we see the nonzero coefficient at 1 Hz and 15 Hz.

As far as the magnitude is concerned, let's work through the DFT of $x[n] = \cos(2\pi n T_s)$. We are looking in particular at the case where we have an integer number of periods. This means that if M is the number of samples in one period, then the total number of data points N is an integer multiple of M, $N = mM$. In this case, $T_s = m/N$. With these preliminaries out of the way, let's take a look at the DFT of $x[n]$

$$
\begin{aligned}
X[k] &= \sum_{n=0}^{N-1} x[n]e^{-jkn\frac{2\pi}{N}} \\
&= \sum_{n=0}^{N-1} \cos(2\pi nT)e^{-jkn\frac{2\pi}{N}} \\
&= \sum_{n=0}^{N-1} \frac{1}{2}\left[e^{j2\pi nT} + e^{-j(2\pi nT}\right] e^{-jkn\frac{2\pi}{N}} \\
&= \sum_{n=0}^{N-1} \frac{1}{2}\left[e^{j2\pi n\frac{m}{N}} + e^{-j(2\pi n\frac{m}{N}}\right] e^{-jkn\frac{2\pi}{N}} \\
&= \frac{1}{2}\sum_{n=0}^{N-1} e^{j\frac{2\pi n(m-k)}{N}} + \frac{1}{2}\sum_{n=0}^{N-1} e^{-j\frac{2\pi n(m+k)}{N}}
\end{aligned}
$$

The first term in this equation corresponds to the positive frequency coefficient, so let's focus on that. This is a summation we encountered earlier when we first started discussing the DFT. We found that

$$\sum_{n=0}^{N-1} e^{jk(2\pi/N)n} = \begin{cases} N & k = 0, \pm N, \pm 2N, \ldots \\ 0 & \text{otherwise} \end{cases}$$

In this case, in the exponent we have $m - k$ instead of k, therefore, the first term above becomes

$$\frac{1}{2}\sum_{n=0}^{N-1} e^{j\frac{2\pi n(m-k)}{N}} = \begin{cases} \frac{N}{2} & k = m, m \pm N, m \pm 2N, \ldots \\ 0 & \text{otherwise} \end{cases}$$

In our particular example, $N = 32$ and $m = 2$, therefore, $X[2] = 16$.

So, what if the data did not contain an integer number of periods? Let's take our program and set $N = 40$, keeping all other parameters the same. The result is shown in Fig. 2.15.

Notice that, even though we are still taking the DFT of a 1 Hz signal, the DFT now contains many nonzero frequency components. There are several ways to explain this. Recall that when we sampled in the frequency domain, we implicitly replicated the signal in the time domain. If we replicate the original signal with an integer number of periods, we would obtain another signal with an integer number of periods, reflecting a signal with the same frequency. However, if we replicate the signal with a non-integer number of periods, we

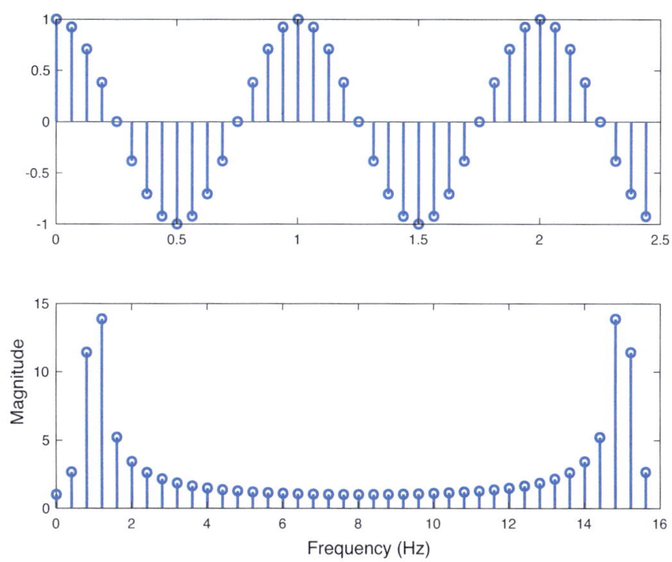

Fig. 2.15 FFT of $\cos(2\pi t)$ sampled at 16 samples per second using 40 samples

introduce other frequencies. You can see this in Fig. 2.16, where we have plotted three repetitions of the signal whose DFT is shown in Fig. 2.15.

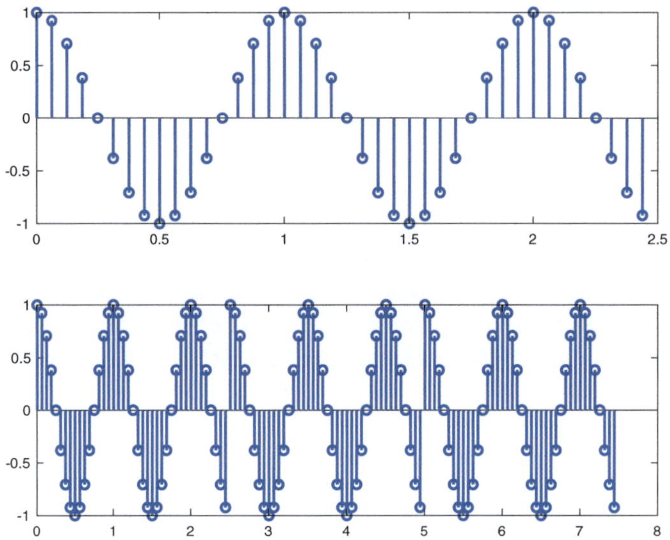

Fig. 2.16 The signal with a non-integer number of periods repeated three times

A somewhat more useful explanation—because it can be applied to signals other than sinusoids—is that the signal we are taking the transform of is not really $\cos(2\pi t)$ because $\cos(2\pi t)$ goes on forever and ever. The signal we are taking the transform of starts at $t = 0$ and goes to zero at $t = 2$ or $t = 2.5$. To mathematically model this signal let's define $r_T(t)$ as follows:

$$r_T(t) = \begin{cases} 1 & 0 \le t \le T \\ 0 & \text{otherwise} \end{cases}$$

then the signal, which contained an integer number of periods

$$x(t) = \cos(2\pi t) \cdot r_2(t)$$

The Fourier transform of $x(t)$ is not the Fourier transform of $\cos(2\pi t)$. Because we are taking the Fourier transform of $\cos(2\pi t) \cdot r_2(t)$, the result will be the Fourier transform of $\cos(2\pi t)$ convolved with the Fourier transform of $r_2(t)$. The Fourier transform $R_2(\omega)$ of $r_2(t)$ can be found as:

$$
\begin{aligned}
\mathcal{F}\left[r_2(t)\right] &= \int_{-\infty}^{\infty} r_2(t) e^{-j\omega t}\, dt \\
&= \int_0^2 e^{-j\omega t}\, dt \\
&= -\frac{1}{j\omega}\left[e^{-j2\omega} - 1\right] \\
&= \frac{1}{j\omega}\left[1 - e^{-j2\omega}\right] \\
&= \frac{2e^{-j\omega}}{\omega}\left[\frac{e^{j\omega} - e^{-j\omega}}{2j}\right] \\
&= \frac{2e^{-j\omega}\sin(\omega)}{\omega}
\end{aligned}
$$

So, when we take the DFT, we should be seeing a sampled version of $2e^{-j2\pi(f-1)}$ $\sin(2\pi(f-1))/2\pi(f-1)$. The DFT samples the DTFT every $2\pi/NT$ radians or $1/NT$ Hz. The number of samples N in this case is 32, while the value of T is 1/16. So the transform is sampled every $\Delta f = 16/32 = 0.5$ Hz. Noting that

$$
\left|\frac{2e^{-j2\pi(f-1)}\sin(2\pi(f-1)}{2\pi(f-1)}\right| = \left|\frac{\sin(2\pi(f-1)}{2\pi(f-1)}\right|
$$

let's look at the magnitude of the samples

f	$\left\lvert\frac{\sin(2\pi(f-1)}{2\pi(f-1)}\right\rvert$
0	$\frac{\sin(-2\pi)}{-2\pi} = 0$
0.5	$\frac{\sin(-\pi)}{-\pi} = 0$
1.0	$\frac{\sin(0)}{0} = 1$ (using l'Hopital's rule)
1.5	$\frac{\sin(\pi)}{\pi} = 0$
2.0	$\frac{\sin(2\pi)}{2\pi} = 0$
$\vdots$	$\vdots$

Let's perform the same calculations for $N = 40$. Instead of multiplying the cosine function with $r_2(t)$, we multiply it with $r_{2.5}(t)$. The Fourier transform of $r_{2.5}(t)$ is given by

$$
\begin{aligned}
\mathcal{F}\left[r_{2.5}(t)\right] &= \int_{-\infty}^{\infty} r_{2.5}(t)e^{-j\omega t}\,dt \\
&= \int_{0}^{2.5} e^{-j\omega t}\,dt \\
&= -\frac{1}{j\omega}\left[e^{-j2.5\omega} - 1\right] \\
&= \frac{1}{j\omega}\left[1 - e^{-j2.5\omega}\right] \\
&= \frac{2e^{-j\frac{5}{4}\omega}}{\omega}\left[\frac{e^{j\frac{5}{4}\omega} - e^{-j\frac{5}{4}\omega}}{2j}\right] \\
&= \frac{2e^{-j\frac{5}{4}\omega}\sin(\frac{5}{4}\omega)}{\omega}
\end{aligned}
$$

In this case $\Delta f = 16/40 = 0.4$.
Evaluating $\left|\frac{\sin(2.5\pi(f-1)}{2\pi(f-1)}\right|$ we obtain.

f	$\left\lvert\frac{\sin(2.5\pi(f-1)}{2\pi(f-1)}\right\rvert$
0	$\left\lvert\frac{\sin(-2.5\pi)}{-2\pi}\right\rvert = 0.16$
0.4	$\left\lvert\frac{\sin(-1.5\pi)}{-1.2\pi}\right\rvert = 0.27$
0.8	$\left\lvert\frac{\sin(-0.5\pi}{-0.4\pi}\right\rvert = 0.8$
1.2	$\left\lvert\frac{\sin(0.5\pi}{0.4\pi}\right\rvert = 0.8$
1.6	$\left\lvert\frac{\sin(1.5\pi)}{1.2\pi}\right\rvert = 0.27$
2.0	$\left\lvert\frac{\sin(2.5\pi)}{2\pi}\right\rvert = 0.16$
$\vdots$	$\vdots$

How do we handle this issue? There are two ways we can somewhat ameliorate this problem. First, noticing that we are not actually taking the Fourier transform of the signal $x(t)$, we are taking the Fourier transform of the signal $x(t)$ times $r_T(t)$, we can replace $r_T(t)$ with some other function that introduces less distortion. These other functions are called windows. Using this terminology, $r_T(t)$ is a rectangular window. Instead of implicitly multiplying the signal $x(t)$ by a rectangular window, we can explicitly multiply the signal by a window of our choice. There are several windows available, including the Hanning window, the Hamming window, the Hann window, the Parzen window, the Blackman-Harris window, and many others. For example, if we multiply our signal with the Hann window given by

$$
w[n] = \sin^2\left(\frac{\pi n}{N}\right)
$$

we get the result shown in Fig. 2.17.

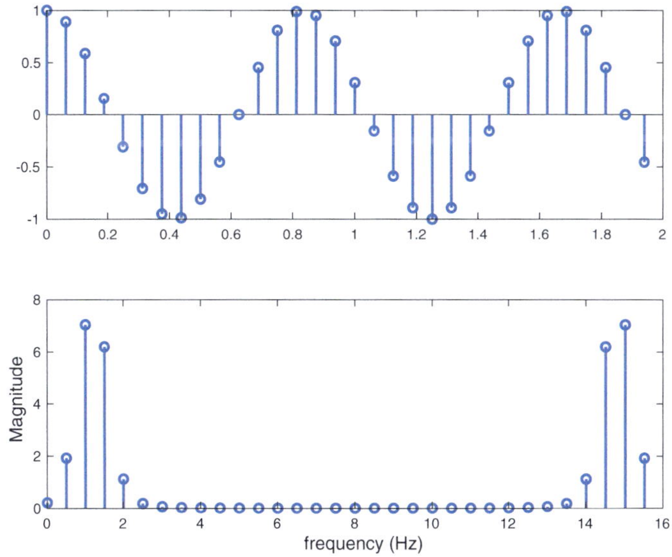

Fig. 2.17 FFT of $\cos(2\pi t)$ sampled at 16 samples per second and windowed using a Hann window using 40 samples

Comparing this to Fig. 2.15, we can see that a lot of the noise in the spectrum—the leakage of the FFT—has been removed. We still have two prominent peaks where we should have only one, but this is primarily due to the way we are sampling the frequency axis. We are sampling the frequency axis at 0.4 Hz intervals, so we never capture the 1 Hz signal. Instead, the two neighboring frequencies at 0.8 Hz and 1.2 Hz show large magnitudes.

If we did not want to use a window, there is one other way we can improve the situation. This is called zero padding. In this, we append a sequence of zeros to the signal. Typically, we select the number of zeros to ensure the total length of the sequence is a power of two. By appending the zeros, we reduce the sampling interval in the frequency domain. If we have some idea of the particular frequency we are interested in, we can set the total length of the sequence to a value that will result in the frequency of interest being one of the samples. In our example, the frequency we are interested in seeing is 1 Hz. If we append enough zeros to the input sequence so that the length of the sequence is 64, the interval between frequencies

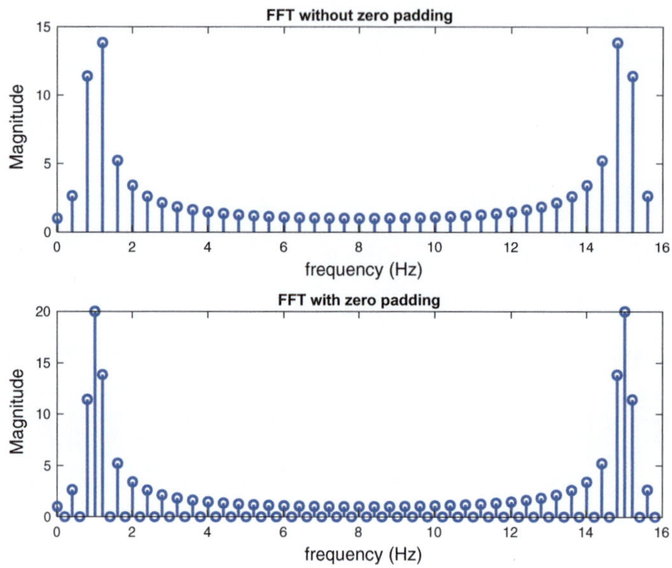

Fig. 2.18 FFT of $\cos(2\pi t)$ sampled at 16 samples per second and windowed using a Hann window using 40 samples

Δf will be $16/64$ or 0.25. As 1 is an integer multiple of 0.25, it should be among the samples of the frequency spectrum. The result of zero padding is shown in Fig. 2.18. Although the leakage remains, the frequency component at 1 Hz is clearly visible.

2.4 A MATLAB Example

Let's take a signal and find its DFT using different-length FFTs in MATLAB. The signal consists of two sinusoids with frequencies f_0 and f_1 sampled at $sampling\_frequency$ samples per second. The total number of data points available to us is N. We will obtain three different transforms: an N-point transform, an $N/2$-point transform, and a $2N$-point transform. For the last transform, we will have to use zero padding. The goal is to see the effect of changing the size of the FFT on the spectral representation and also to elucidate some of the peculiarities of MATLAB.

We begin by defining our variables

```
clear all
sampling_frequency = 100;
N = 200;
fo = 15; % frequency of the first sinusoid
f1 = 26; % frequency of the second sinusoid\
```

Notice the first line. This clears all the variables from memory. MATLAB retains data in memory, which can sometimes complicate matters. Suppose you ran a program with an array x that had certain properties. If you don't use the *clear all* command and you define a variable x, unbeknownst to you, this 'new' variable will have all the properties/values that the 'old' variable had. Putting *clear all* as the first line in your MATLAB program is useful.

Since we will eventually need an array of size $2N$, let's define our data array x to be of that size. We can do that—and fill the array with zeros by using the *zeros* command in MATLAB.

```
x = zeros(1,2*N);  % This creates a 1 X 2*N array filled with zeros
```

The time points at which the sinusoids will be sampled are going to be $1/sampling\_frequency$, and there will be N of these samples. So the time points for these samples can be obtained by

```
t = [0:1:N-1]/sampling_frequency; % t goes from 0 to N/sampling_frequency in steps
                                  % of 1/sampling_frequency
```

Now we can generate the data

```
x = cos(2*pi*fo*t) + cos(2*pi*f1*t);
```

Because the array t contains N values, the first N values of x will be filled with the sampled values of the sinusoids. We now take the N point FFT of x and plot x and the absolute value of the FFT.

```
figure
subplot(4,1,1)
plot(t,x(1:1:N))
y1 = fft(x,N); %The fft is the same size as the input
faxis1 = (0:sampling_frequency/N:sampling_frequency-.5); % Notice the fudging you have to
                                                          % do to make the arrays the same
                                                          % length for the plot command. In
                                                          % this case I subtracted 0.5 from
                                                          % the upper limit to make both
                                                          % arrays the same size
subplot(4,1,2)
plot(faxis1,abs(y1));
```

Notice the frequency axis. The interval between frequency samples is given by $sampling\_frequency/N$, and the range of frequencies is from 0 to the sampling frequency. Or at least that is what it should be. If we try to do that, we find that the frequency axis array has one extra point. It is of size 201, where the data array is of size 200. For the plot command to work correctly, we ended up reducing the upper limit by $sampling\_frequency/N$. This is not the only way to fix this problem. We could have replaced $faxis1$ in the plot command with $faxis1(1 : N)$.

Then we repeat the process to obtain the $N/2$-point FFT. This time, instead of fudging the upper limit of the frequency axis array, we used $faxis2(1 : N/2)$ in the plot command.

```
%  Now let's take the FFT of half the data values
y2 = fft(x,N/2);
faxis2 = (0:sampling_frequency/(N/2):sampling_frequency);
subplot(4,1,3)
plot(faxis2(1:N/2),abs(y2)) % This time instead of fudging the upper limit we simply
                            % made sure that the frequency axis array
                            % faxis2 had the same number of points as the
                            % data array
```

Finally, we take the $2 * N$ point FFT. Recall that the zero padding has already been taken care of when we initialized the data array to be an array of size $2N$ filled with zeros. The first N values are now samples of the sinusoids, but the last N values are still zero. The fft function in MATLAB automatically takes care of this, but the FFT program in other environments may not. It doesn't hurt to be safe.

```
% remember that x was initialized to 2*N zeros before we
% put the samples of cosine in the first N locations

y3 = fft(x,2*N);
faxis3 = (0:sampling_frequency/(2*N):sampling_frequency);
subplot(4,1,4)
stem(faxis3(1:2*N),abs(y3))
```

The results are shown in Fig. 2.19, where the first plot displays the data, and the second plot shows the magnitudes of the frequency components generated using an N- point FFT. The third plot is the FFT of the first $N/2$ values, and the last plot is the result of zero padding with N zeros.

Notice that in the first two cases, we see only one nonzero frequency component, and that is precisely the frequency component present in the signal. In the third case, however, you see other nonzero positive values. The reason for this is that what we actually have in the frequency domain are two $\sin(x)/x$ functions centered at the frequencies of the sinusoids. These sinusoids have been sampled, and it just so happens that the samples always land on the zero crossings of the $\sin(x)/x$ functions. In the last case, because of the zero padding, the samples are closer together, so we get not only the samples of the zero crossings but also of the other parts of the $\sin(x)/x$ function as well.

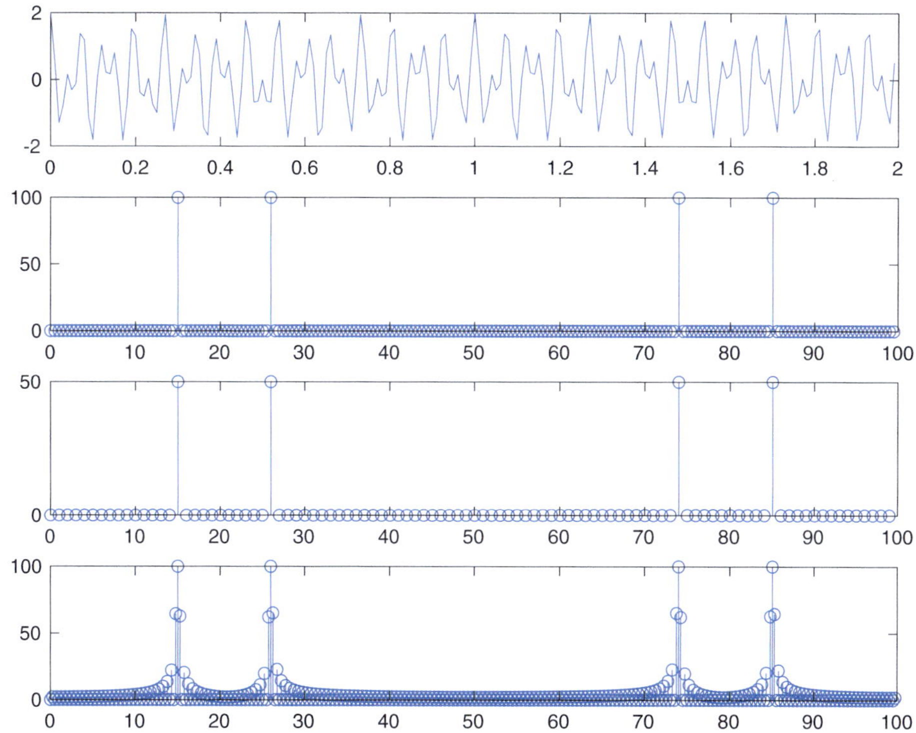

Fig. 2.19 Plot of the signal and the magnitude of the three FFTs. The top one is the N point FFT, the next plot is for the $N/2$ point FFT, and the last plot is for the $2N$ point FFT

We can see this effect more clearly if we pick sinusoids with non-integer frequencies. Let's pick $f_0 = 15.4$ and $f_2 = 26.2$. The resulting FFT magnitudes are shown in Fig. 2.20.

2.5 Summary

In this chapter, we have examined how we represent the signals that we will later process. Overwhelmingly, the signals of interest to us are analog. Overwhelmingly, our processing of these signals is in the digital domain. Therefore, we need to obtain digital representations of these analog signals. This digital representation of analog signals entails compromises that we should be aware of. For example, when we are given EEG signals sampled at 256 samples per second, we should know that if the signal had originally contained information at frequencies close to or above 128 Hz, that information is lost. When examining the spectral behavior of signals, it is essential to understand the effect of the assumed periodicity on the view presented to us. When examining eight-bit samples of a signal, it is vital to be aware that the number of bits used limits the accuracy of those samples. If we increased the number

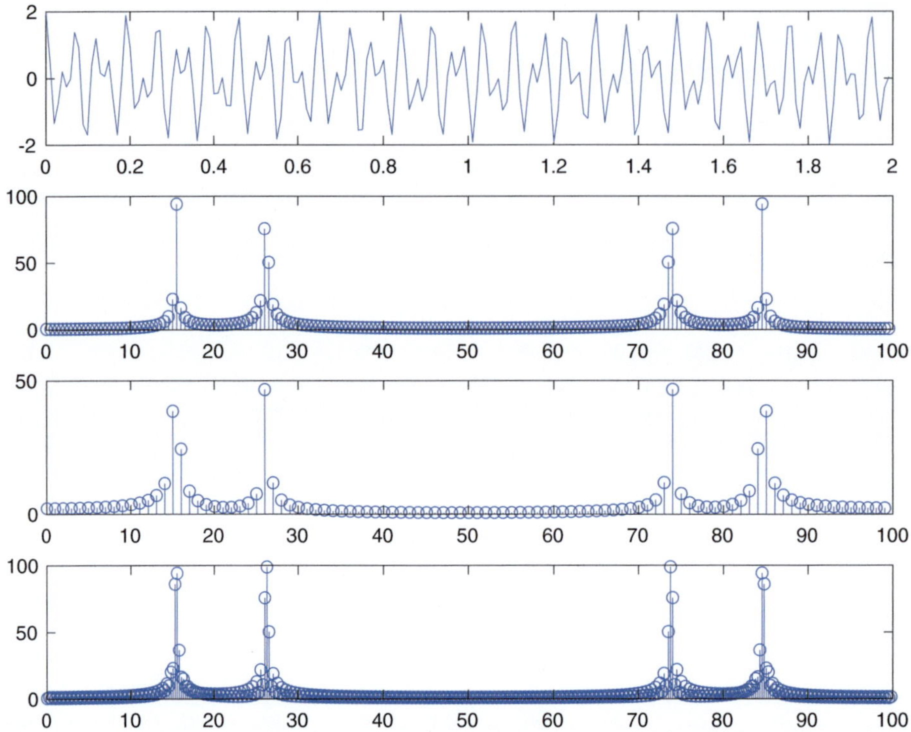

Fig. 2.20 Plot of the signal and the magnitude of the three FFTs. The top one is the N point FFT, the next plot is for the $N/2$ point FFT, and the last plot is for the $2N$ point FFT

of bits from eight to sixteen, we would significantly reduce the distortion in our samples. Whether this reduction in distortion is worth the increase in the number of bits we need to process and store depends on the particular application.

The representation of a signal is simply that—a representation. It is not the signal itself. All kinds of distortions may creep into our representations; the more we know about them, the more we will understand the actual signal behind the representation.

If you wish to explore this subject further, there are several excellent books that cover the topic in greater detail. A classic book on the subjects covered in this chapter is the book by Oppenheim et al. [5]. A more concise coverage can be found in [6]. There is also a series of videos covering these topics [7].

Discrete Time Filters

3

The Discrete Fourier Transform (DFT) provides a signal representation in terms of all the frequencies present in the signal. Often, however, we are interested only in certain frequencies or in a range of frequencies. Discrete-time filters can provide a simple and computationally efficient way of isolating ranges of frequencies. Consider, for example, the case where we want to remove the DC value from a signal. We could take the DFT of the signal, replace the zero-frequency component with zero, and then take the inverse DFT. A much simpler route would be to implement the filter with an input/output relationship

$$y[n] = x[n] - x[n-1]$$

where $\{x[n]\}$ are the input to the filter and $\{y[n]\}$ are the output of the filter. Similarly, to smooth out high-frequency variation in the signal, we could use an averaging filter with an input/output relationship

$$y[n] = \frac{x[n] + x[n-1]}{2}$$

While simple to implement, each of these filters, in addition to accomplishing the intended task, also introduces some unwanted effects. The differencing filter not only removes the DC value of the signal, but it can also significantly attenuate some low-frequency components in the signal we might wish to keep. Similarly, the averaging filter lacks specificity regarding the range of frequencies that are attenuated. To obtain filters that can be more targeted to particular objectives, we will need some understanding of discrete-time filters. The goal of this chapter is to develop this understanding. The goal is not to become experts in designing digital filters. There are numerous packages available for designing digital filters. Instead, we aim to acquire enough knowledge to be informed users. The Z transform can help us with this understanding. While a complete exposition of the Z-transform is beyond our scope, we can briefly review some aspects of the Z-transform that will be useful to us.

K. Sayood, *Biomedical Signal Processing*, Synthesis Lectures on Biomedical Engineering,
https://doi.org/10.1007/978-3-032-08855-0_3

3.1 A Brief Review of Z Transforms

The Z transform of a sequence $\{x[n]\}$ is defined as

$$X(z) = \sum_{n=-\infty}^{\infty} x[n]z^{-n}$$

where $z = re^{j\omega}$ and ω is the frequency in radians. To find the frequency representation of the discrete time signal $x[n]$ we set $r = 1$ in $z = re^{j\omega}$ and we get

$$X(e^{j\omega}) = \sum_{n=-\infty}^{\infty} x[n]e^{-jn\omega}$$

which is the expression for the discrete-time Fourier transform. As in the case of the DTFT, we vary ω from 0 to 2π. If $\{x[n]\}$ are discrete time samples of a continuous function $x(t)$ then the Fourier transform of $\{x[n]\}$ is simply a periodic repetition of the Fourier transform of the $x(t)$ where the repetition period is $2\pi/T_s$ and T_s is the interval between samples. As we did for the discrete-time Fourier transform, we have normalized the sampling interval to $T_s = 1$, so the frequency representation is periodic with period 2π. If we want to go back from these normalized frequencies to the "actual" frequencies, all we need to do is to map the sampling frequency to 2π, and interpret the values between 0 and 2π as the frequencies between 0 and the sampling frequency. We will see some examples later.

To use the Z transform to study discrete-time filters, we need to use a few of the properties of the Z transform. The first and most important is the convolution property, which states that the Z-transform of the convolution of two sequences is the product of the Z-transforms of the individual sequences.

$$\mathcal{Z}\left[a[n] \circledast b[n]\right] = A(z)B(z)$$

where $A(z)$ is the Z transform of the sequence $\{a[n]\}$ and $B(z)$ is the Z transform of the sequence $\{b[n]\}$.

The output of a linear discrete-time system is given by the convolution of the input and the impulse response of the system

$$y[n] = x[n] \circledast h[n]$$

therefore, in the Z domain

$$Y(z) = X(z)H(z)$$

where $H(z)$ is the transfer function of the discrete-time filter. For the discrete-time filters of interest to us, the transfer function will be a rational polynomial in z

$$H(z) = \frac{N(z)}{D(z)}$$

where $N(z)$ is the numerator polynomial and $D(z)$ is the denominator polynomial. We can find the roots of these polynomials and write the transfer function as

$$H(z) = \frac{K(z - z_0)(z - z_1)\ldots(z - z_m)}{(z - p_0)(z - p_1)\ldots(z - p_n)}$$

where $\{z_i\}$ are called the *zeros* of the system and $\{p_i\}$ are called the *poles* of the system. There is a good reason for the names—when z takes on a value equal to any of the zeros of the system the transfer function $H(z)$ becomes zero, and when z is equal to one of the poles p_i the denominator polynomial takes on a value of zero driving the value of $H(z)$ to be infinity. When z takes on a value close to one of the zeros, the effect is to reduce the magnitude of the numerator and hence of $H(z)$. Similarly, when z takes on a value close to one of the poles, this reduces the magnitude of the denominator, which in turn results in an increase in the magnitude of $H(z)$. In these situations, rather than $H(z)$ going to zero or infinity, the magnitude of $H(z)$ decreases or increases. As in the case of the Laplace transform, the location of the poles in the Z transform determines the stability of the system. For causal systems, the requirement for a system to be stable is that the poles all lie within the unit circle.

One of the advantages of the Z transform is that it allows us to go directly from the transfer function to the difference equation that relates the input and output. To do this, we will need the following property of the Z transform, which shows the Z transform of a delayed signal:

$$\mathcal{Z}\left[x[n - n_o]\right] = z^{-n_o}X(z)$$

To use this property to directly obtain the difference equation, we first write $H(z)$ as a rational polynomial in z.

$$H(z) = \frac{b_0 z^M + b_1 z^{M-1} + b_2 z^{M-2} + \cdots + b_M}{z^N + a_1 z^{N-1} + a_2 z^{N-2} + \cdots + a_N}$$

In most cases of interest to us, M will be equal to N. Let's assume that to be the case and divide top and bottom by z^M (or z^N), then

$$H(z) = \frac{b_0 + b_1 z^{-1} + b_2 z^{-2} + \ldots b_M z^{-M}}{1 + a_1 z^{-1} + a_2 z^{-2} + \cdots + a_N z^{-N}}$$

Noting that

$$Y(z) = X(z)H(z)$$

we get

$$Y(z) = X(z) \cdot \left[\frac{b_0 + b_1 z^{-1} + b_2 z^{-2} + \ldots b_M z^{-M}}{1 + a_1 z^{-1} + a_2 z^{-2} + \cdots + a_N z^{-N}}\right]$$

or

$$Y(z)\left[1+a_1z^{-1}+a_2z^{-2}+\cdots+a_Nz^{-N}\right]=X(z)\left[b_0+b_1z^{-1}+b_2z^{-2}+\ldots b_Mz^{-M}\right]$$

Multiplying through, we get

$$Y(z)+a_1z^{-1}Y(z)+a_2z^{-2}Y(z)+\cdots+a_Nz^{-N}Y(z)=b_0X(z)+b_1z^{-1}X(z)+b_2z^{-2}X(z)+\ldots b_Mz^{-M}X(z)$$

To get the time domain relationship between the input and output, we take the inverse transform of both sides of the equation. The linearity of the Z transform means that the inverse Z transform of the left side of the equation is given by the sum of the inverse Z transform of the individual terms on the left-hand side.

$$\mathcal{Z}^{-1}\left[Y(z)+a_1z^{-1}Y(z)+a_2z^{-2}Y(z)+\cdots+a_Nz^{-N}Y(z)\right]=$$
$$\mathcal{Z}^{-1}\left[Y(z)\right]+\mathcal{Z}^{-1}\left[a_1z^{-1}Y(z)\right]+\mathcal{Z}^{-1}\left[a_2z^{-2}Y(z)\right]+\cdots+\mathcal{Z}^{-1}\left[a_Nz^{-N}Y(z)\right]$$

Using the fact that

$$\mathcal{Z}^{-1}\left[z^{-n_o}Y(z)\right]=y[n-n_o]$$

the inverse transform of the left side of the equation becomes

$$y[n]+a_1y[n-1]+a_2y[n-2]+\cdots+a_Ny[n-N]$$

Similarly, the inverse transform of the right side of the equation becomes

$$b_0x[n]+b_1x[n-1]+b_2x[n-2]+\ldots b_Mx[n-M]$$

and we get the relationship

$$y[n]+a_1y[n-1]+a_2y[n-2]+\cdots+a_Ny[n-N]=b_0x[n]+b_1x[n-1]+b_2x[n-2]+\ldots b_Mx[n-M]$$

From this, we get the input/output relationship of a linear discrete-time filter.

$$y[n]=\sum_{i=0}^{M}b_ix[n-i]-\sum_{i=1}^{N}a_iy[n-i]$$

The set of values $\{a_i, b_i\}$ is referred to as the filter coefficients. By selecting different values for the coefficients, we can obtain various discrete-time filters. In most applications, we want these coefficients to be real. For this to happen, the poles and zeros have to be either real or occur in complex conjugate pairs.

3.2 Some Simple Filters

With that rather quick review of Z transforms under our belt, let's make this a bit more concrete. Let's suppose our filter is the simple filter we described to get rid of the DC value:

$$y[n] = x[n] - x[n-1]$$

To find the transfer function, let's take the Z transform of both sides of the equation

$$Y(z) = X(z) - z^{-1}X(z) = X(z)\left[1 - z^{-1}\right]$$

from which we get

$$H(z) = \frac{Y(z)}{X(z)} = 1 - z^{-1} = \frac{z-1}{z}$$

This transfer function has a pole at the origin and a zero at $z = 1$ as shown in Fig. 3.1.

To obtain the frequency response of this filter, we set $r = 1$ in $z = re^{j\omega}$ and evaluate the function from $\omega = 0$ to π. Another way of saying this is that we evaluate $H(z)$ on the unit circle in the z-plane, where the unit circle is a circle of radius 1. If we move around the unit circle to find the transfer function of this filter, we see that the pole at the origin is equidistant to all points on the unit circle. Therefore, the pole affects all frequencies equally. The zero occurs at $z = 1$, or $\omega = 0$ (remember $z = re^{j\omega}$). This will cause the transfer function at $\omega = 0$ to be zero. Therefore, the frequency component of the input at $\omega = 0$, or the DC component, will get multiplied by zero. We have plotted the magnitude of this filter in Fig. 3.2. As can be seen from the figure, this filter will block the DC value but will also attenuate the low-frequency components of the input signal.

What if we wanted the filter to zero out some other frequency component? For example, in many biomedical applications, it is often necessary to eliminate the effect of the 60 Hz line current. To zero out this component, we need to place a zero at the point on the unit circle corresponding to 60 Hz. To find this point, we need to know the sampling frequency used to

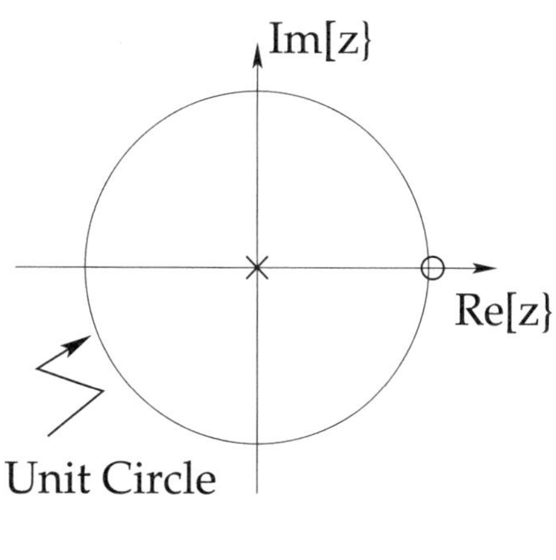

Fig. 3.1 Pole zero pattern for $H(z) = 1 - z^{-1}$

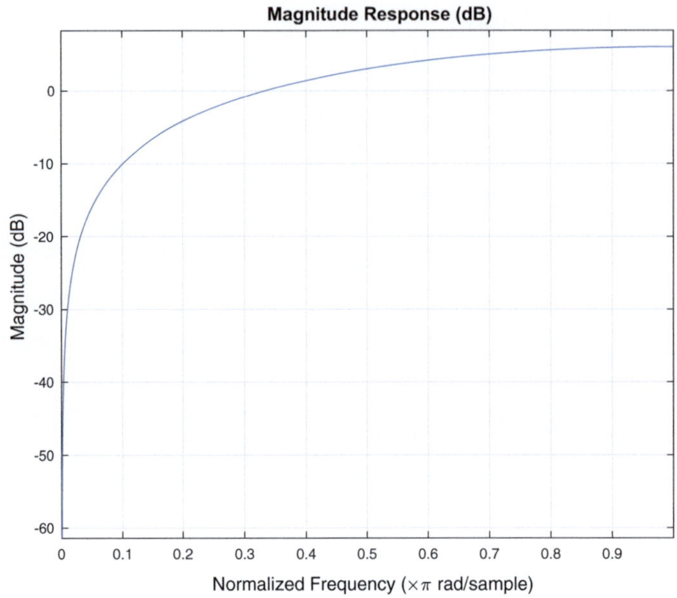

Fig. 3.2 Magnitude of the frequency response of the filter with transfer function $H(z) = 1 - z^{-1}$

generate the signal of interest. Recall that $\omega = 2\pi$ corresponds to the sampling frequency ω_s. Therefore, 60 Hz will map to $\omega_1 = [(2\pi \times 60)/\omega_s] \times 2\pi$. Suppose our signal was sampled at 240 samples per second. This means we would put one zero at $\omega_1 = [(2\pi \times 60)/(2\pi \times 240)] \times 2\pi = \pi/2$. Because this is a complex-valued zero, we also need a zero at $\omega = -\pi/2$ for the coefficients to be real-valued. To offset the zeros, we will put two poles at the origin to give us the pole-zero pattern shown in Fig. 3.3.

Fig. 3.3 Pole zero pattern for $H(z) = \frac{z^2+1}{z^2}$

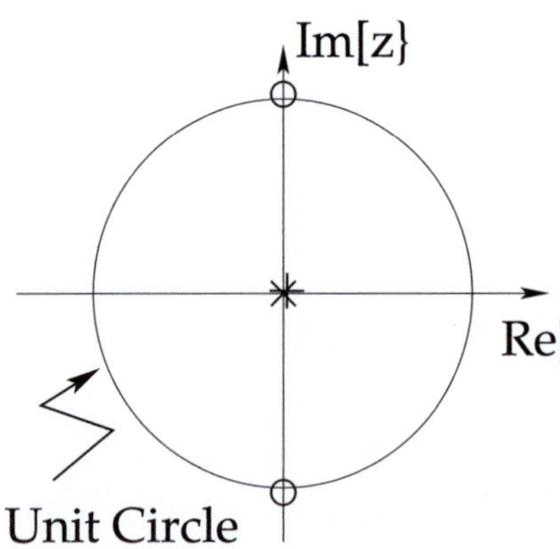

The zeros are at $z = e^{j\frac{\pi}{2}}$ and $z = e^{-j\frac{\pi}{2}}$, and the poles are at $z = 0$. Therefore, the transfer function is given by

$$H(z) = \frac{(z - e^{j\frac{\pi}{2}})(z - e^{-j\frac{\pi}{2}})}{(z - 0)(z - 0)} = \frac{(z - j)(z + j)}{z^2} = \frac{z^2 + 1}{z^2} = 1 + z^{-2}$$

where we have used the fact that

$$e^{j\frac{\pi}{2}} = \cos\left(\frac{\pi}{2}\right) + j\sin\left(\frac{\pi}{2}\right) = 0 + j = j$$

and

$$e^{-j\frac{\pi}{2}} = \cos\left(\frac{\pi}{2}\right) - j\sin\left(\frac{\pi}{2}\right) = 0 - j = -j$$

Noting that $Y(z) = X(z)H(z)$, we substitute for $H(z)$ to obtain

$$Y(z) = X(z)\left[1 + z^{-2}\right] = X(z) + z^{-2}X(z)$$

Now let's take the inverse Z transform of both sides of this equation. We do that by taking the inverse Z transform of each term

$$\mathcal{Z}^{-1}\left[Y(z)\right] = y[n]$$
$$\mathcal{Z}^{-1}\left[X(z)\right] = x[n]$$
$$\mathcal{Z}^{-1}\left[z^{-2}X(z)\right] = x[n-2]$$

Then we use the fact that the Z transform is a linear transform to put this all together and obtain the input-output relationship for this filter

$$y[n] = x[n] + x[n-2]$$

In this case, the sampling frequency was such that the zeros were conveniently at $\pm j$. Generally, we won't be so fortunate. Let's repeat the exercise using a generic sampling frequency f_s and design a filter to reject the frequency component at f_o Hz. In this general case, we would put the zero at an angle of $\omega_1 = 2\pi[(2\pi f_o)/(2\pi f_s)] = 2\pi(f_o/f_s)$. Define

$$z_1 = e^{j\omega_1}$$

then the transfer function becomes

$$H(z) = \frac{(z - z_1)(z - z_1^*)}{z^2} = \frac{z^2 - z(z_1 + z_1^*) + |z_1|^2}{z^2}$$

For any complex number z_1, $z_1 + z_1^* = 2Re[z_1]$, the sum of a number and its complex conjugate is twice the real part of the complex number. Therefore,

$$H(z) = \frac{z^2 - 2Re[z_1]z + 1}{z^2}$$

Dividing top and bottom by z^2 we get

$$H(z) = 1 - 2Re[z_1]z^{-1} + z^{-2}$$

The output $Y(z)$ is given by

$$Y(z) = H(z)X(z) = (1 - 2Re[z_1]z^{-1} + z^{-2})X(z) = X(z) - 2Re[z_1]z^{-1}X(z) + z^{-2}X(z)$$

from which we can obtain the input-output relationship for the filter as

$$y[n] = x[n] - 2Re[z_1]x[n-1] + x[n-2]$$

To see whether this simple filter is sufficient to remove a 60 Hz interference in an EKG signal, let's take the EKG signal shown in Fig. 3.4. If we add a 60 Hz interference signal we get the signal shown in Fig. 3.5. The amount of interfering signal is small but visible.

Applying the filter described above we get the output shown in Fig. 3.6. Clearly, the filter did not perform very well in terms of recovering the original EKG signal. Instead, it seems to have added more distortion. To understand why, let's examine the frequency response of the filter. We can obtain the frequency response directly by evaluating the filter transfer function $H(z)$ on the unit circle, where z is replaced with $e^{j\omega}$, and then calculating the magnitude and

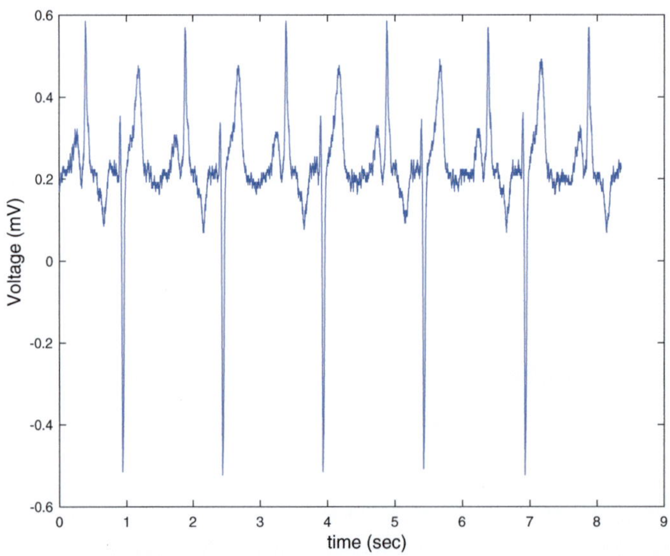

Fig. 3.4 A sample EKG signal

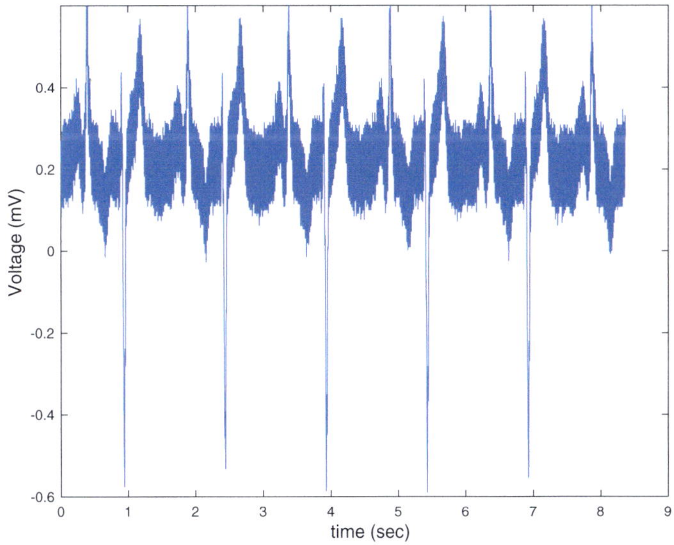

Fig. 3.5 A sample EKG signal with 60 Hz interference

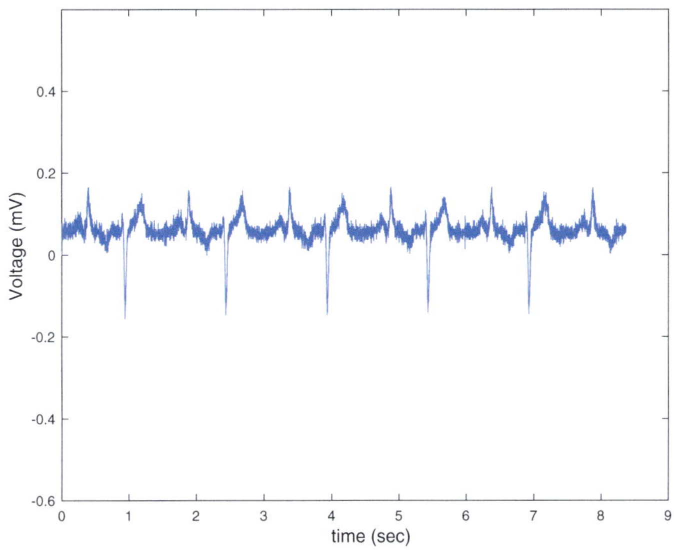

Fig. 3.6 Corrupted EKG signal filtered using the linear filter with input/output relationship $y[n] = x[n] - 2Re[z_1]x[n-1] + x[n-2]$ where z_1 is the location of the pole

phase of the transfer function for different values of the frequency. In MATLAB, the function *freqz* can be used to find the transfer function. Let's suppose that the transfer function is given by

$$H(z) = \frac{b_0 z^M + b_1 z^{M-1} + b_2 z^{M-2} + \cdots + b_M}{a_0 z^N + a_1 z^{N-1} + a_2 z^{N-2} + \cdots + a_N}$$

Let's put the coefficients $\{b_i\}$ and $\{a_i\}$ into arrays b and a. The following MATLAB code will generate the magnitude and phase response of the filter:

```
clear all

noiseFreq = 60; % in Hz
samplingFreq = 720;
numpoints = 600;
% filter the noise by putting zeros on the unit circle
% at the noise frequency
zangle = 2*pi*(noiseFreq/samplingFreq); % angle of the noise frequency
beta = 2*cos(angle);
b = [1 -beta 1];
a = [1];

freqz(b,a,numpoints,samplingFreq)
```

The output of the program is shown in Fig. 3.7.

While this filter effectively eliminates the exact 60 Hz component, it also distorts the signal components that are not at 60 Hz. So what is to be done? The poles in this particular filter have not been doing much. The poles are both located at the origin, equidistant from all points on the unit circle, and so have the same effect at all frequencies. What we could do is move the poles out close to the zeros to cancel out the effect of the zeros as we move away from 60 Hz, as shown in Fig. 3.8.

The transfer function for this filter would be

$$H(z) = \frac{(z - z_1)(z - z_1^*)}{(z - p_1)(z - p_1^*)} = \frac{z^2 - 2Re[z_1]z + 1}{z^2 - 2Re[p_1]z + |p_1|^2}$$

Dividing top and bottom by z^2 we get

$$H(z) = \frac{1 - 2Re[z_1]z^{-1} + z^{-2}}{1 - 2Re[p_1]z^{-1} + |p_1|^2 z^{-2}}$$

Using the fact that $Y(z) = X(z)H(z)$ we get

$$Y(z)\left[1 - 2Re[p_1]z^{-1} + |p_1|^2 z^{-2}\right] = X(z)\left[1 - 2Re[z_1]z^{-1} + z^{-2}\right]$$

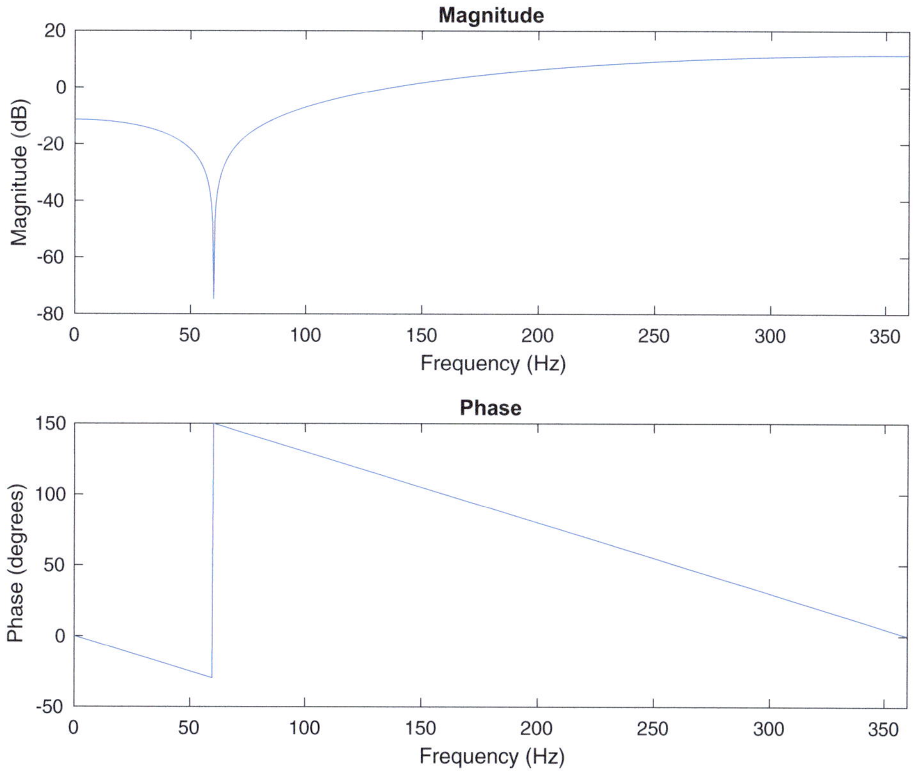

Fig. 3.7 Magnitude and phase of the simple filter with poles at the origin and zeros at the frequency to be cancelled

Fig. 3.8 Moving the poles of the filter from the origin in (**a**) to try and cancel out the effect of the zeros in (**b**)

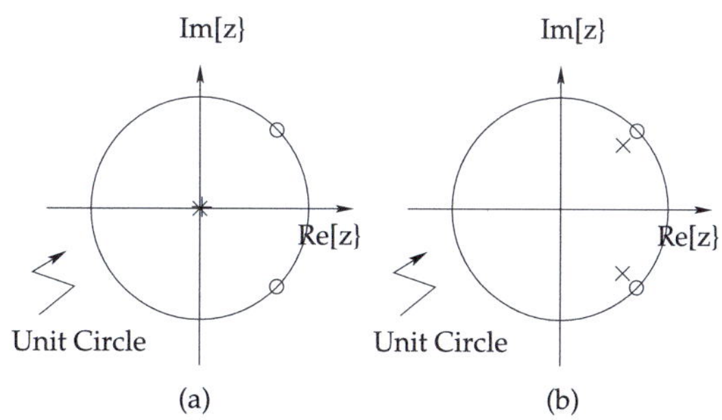

or

$$Y(z) - 2Re[p_1]z^{-1}Y(z) + |p_1|^2 z^{-2}Y(z) = X(z) - 2Re[z_1]z^{-1}X(z) + z^{-2}X(z)$$

Taking the inverse Z transform, using the fact that the inverse Z transform of $z^{-m}Y(z)$ is $y[n-m]$ and the inverse Z transform of $z^{-m}X(z)$ is $x[n-m]$, we get

$$y[n] - 2Re[p_1]y[n-1] + |p_1|^2 y[n-2] = x[n] - 2Re[z_1]x[n-1] + x[n-2]$$

or

$$y[n] = x[n] - 2Re[z_1]x[n-1] + x[n-2] + 2Re[p_1]y[n-1] - |p_1|^2 y[n-2]$$

When we set the magnitude of p_1 to 0.99 the input-output relationship becomes

$$y[n] = x[n] - 1.732x[n-1] + x[n-2] + 1.71468y[n-1] - 0.9801y[n-2]$$

If we now use this to operate on the EKG sequence corrupted by 60 Hz noise, we get the filtered output shown in Fig. 3.9. A significantly better result than we obtained when we left the poles at the origin. There is still some distortion in the early part of the output sequence. This is due to the initial transient response of the filter. But once the filter gets going, it does a pretty good job of removing the 60 Hz distortion. We can also see why this filter works better if we look at the magnitude of the transfer function plotted in Fig. 3.10. Unlike the

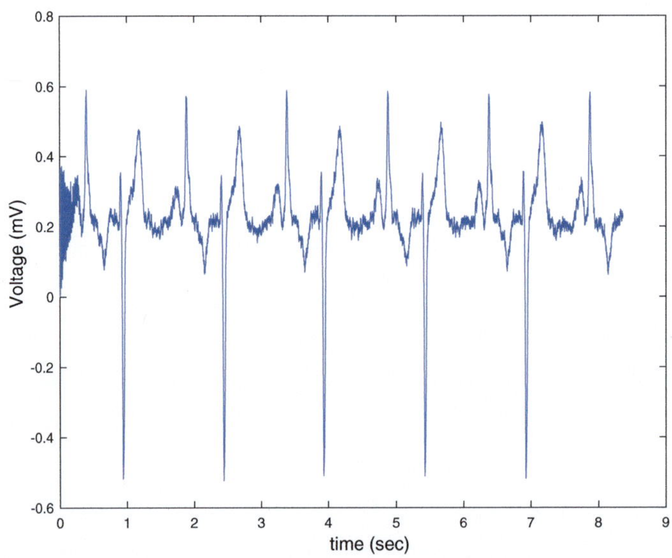

Fig. 3.9 Corrupted EKG signal filtered using the linear filter with input/output relationship $y[n] = x[n] - 2Re[z_1]x[n-1] + x[n-2]$ where z_1 is the location of the pole

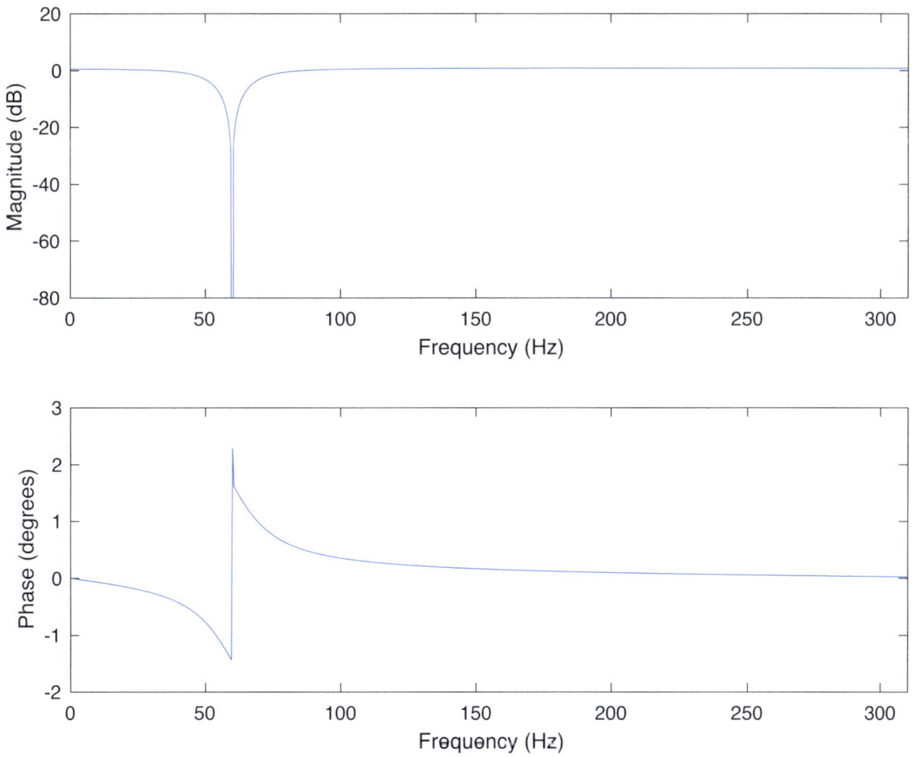

Fig. 3.10 Magnitude and phase of the filter with transfer function $H(z) = \frac{1-2Re[z_1]z^{-1}+z^{-2}}{1-2Re[p_1]z^{-1}+|p_1|^2z^{-2}}$

previous filter, along with the sharp notch at 60 Hz, this filter has a flat passband for all other frequencies.

What is the difference between the two filters? In the first case, the output is entirely determined by a weighted sum of the current and past inputs. In the second case, the current output is determined by a weighted sum of current and past inputs *and* a weighted sum of past outputs. These two filter types are known as finite impulse response (FIR) and infinite impulse response (IIR) filters. There are advantages and disadvantages to both kinds of filters, and, depending on the application, one or the other may be preferable. Let's take a brief look at both types of filters.

FIR filters We begin with FIR filters. Recall that the input-output relationship of a general linear discrete-time filter is given by

$$y[n] = \sum_{i=0}^{M} b_i x[n-i] - \sum_{i=1}^{N} a_i y[n-i]$$

If we set $a_i = 0$ for all i we get

$$y[n] = \sum_{i=0}^{M} b_i x[n-i]$$

To obtain the impulse response of this filter we set $x[n] = \delta[n]$

$$h[n] = \sum_{i=0}^{M} b_i \delta[n-i]$$

Using the fact that

$$\delta[n] = \begin{cases} 1 & n = 0 \\ 0 & n \neq 0 \end{cases}$$

we can see that

$$h[n] = \begin{cases} b_n & n = 0, 1, \ldots, M \\ 0 & \text{otherwise} \end{cases}$$

That is, $h[n]$ is nonzero for only $M + 1$ values. As long as M is finite, the impulse response is nonzero for only a finite number of values. This is why such a filter, whose output is only a function of the input, and not of its past outputs, is called a finite impulse response (FIR) filter.

There are several advantages to this kind of filter. Because the impulse response is absolutely summable,

$$\sum_{n=-\infty}^{\infty} |h[n]| = \sum_{n=0}^{M} |b_n|$$

the FIR filter is always stable. We can also see this from the fact that all the poles of this filter are at the origin and, thus, by definition, within the unit circle. This is particularly useful because, as you saw in the example above, an FIR filter with only a few coefficients is not particularly good. To obtain a decent FIR filter, we usually require a relatively large number of nonzero coefficients. Due to historical reasons, the number of nonzero coefficients in an FIR filter is sometimes referred to as the number of taps in the filter. We can also show that it is easy to obtain FIR filters with linear phase. Why would we want a linear phase? Consider a simple delay circuit in which the output is simply the input delayed by n_o samples

$$y[n] = x[n - n_o]$$

To obtain the transfer function of the filter, take the Z transform of both sides.

$$Y(z) = z^{-n_o} X(z)$$

Then

$$H(z) = \frac{Y(z)}{X(z)} = z^{-n_o}$$

Evaluating the transfer function on the unit circle

$$H(e^{j\omega}) = e^{-j\omega n_o}$$

The magnitude of this filter is clearly one, and the phase is linear

$$\theta(\omega) = -\omega n_o$$

The linearity of the phase allows the entire signal to be delayed the same amount. If the phase were not linear, different frequency components would be delayed by different amounts, resulting in the smearing of the input signal. When a filter has a linear phase, we don't experience this frequency-dependent distortion, making it a particularly desirable property to have.

The disadvantage of the FIR filter is that to achieve decent performance, we need a large number of taps; M has to be relatively large. As the filter coefficients, as well as the past $M - 1$ values of the input, need to be stored, this increases the amount of memory required to implement the filter. Furthermore, the increase in the number of coefficients also increases the number of multiplications required for generating each filter output, thereby increasing the computational complexity of the filter.

IIR Filters If, in our input-output relationship,

$$y[n] = \sum_{i=0}^{M} b_i x[n-i] + \sum_{i=1}^{N} a_i y[n-i]$$

the values of a_i are not zero, then we get a recursive filter—past values of the output are used along with the input to obtain the current output value. This recursion means that the effect of an impulse at $n = 0$ continues forever. Hence, such a filter is referred to as an infinite impulse response (IIR) filter. As we saw from our example, for the same number of coefficients, an IIR filter is much more efficient than an FIR filter. Another way of saying this is that to achieve a set of requirements, we need fewer coefficients if we use an IIR filter than if we use an FIR filter. This means that the implementation of the IIR filter is computationally more efficient, as fewer coefficients result in fewer arithmetic operations. These advantages come at a cost. It is much more challenging to get a linear phase. The fact that there is feedback means that any error in the signal path persists indefinitely. In fact, quantization errors can result in output even when there is no input.

3.3 Design of Filters

In the preceding example, we had a very simple objective—to eliminate a single frequency component—and designing a filter to accomplish this objective was also straightforward. When the objective becomes more complex, such as keeping or blocking a range of frequencies, this approach may become impractical. Fortunately, when faced with such a situation, several tools are available to help us design these more complex filters. Later in this chapter, we will describe some of the tools available in MATLAB. Similar tools are also available for many other environments, including *Octave* and *Python*.

Before digital filters existed, there were analog filters. Many of the design techniques for digital filters work by mapping the requirements for digital filters to the analog domain, using well-known techniques for the design of analog filters to create filters that satisfy these requirements, and then translating this design into the digital domain. In the analog domain, the focus is primarily on designing prototype low-pass filters, which can then be translated into high-pass, band-pass, or band-stop filters.

In the ideal world, a filter would either pass certain frequency components without distortion or block frequency components without any leakage. For example, the magnitude of an ideal low-pass filter would look like the one shown in Fig. 3.11. The magnitude of the passband is completely flat, meaning that the components of the input signal corresponding to the band of frequencies that make up the passband pass through unaltered. The magnitude of the filter in the stopband is zero, meaning that all frequency components of the signal that fall within the stopband are completely blocked. The cutoff frequency f_c is the transition between the passband and the stopband. In practice, the magnitude characteristics of a low-pass filter look more like what is shown in Fig. 3.12.

The cutoff frequency f_c is no longer the boundary between the passband and the stopband. Instead, we usually define the cutoff frequency as the half-power point, or the frequency at which the magnitude of the filter in decibels is 3 dB below the maximum amplitude. The transition between the passband and stopband is not abrupt. Instead, there is a transition band between the passband and the stopband. The magnitude of the filter in the passband is not necessarily constant. Instead, there are deviations in magnitude in the passband, referred to as the passband ripple. There may be similar deviations in the stopband—known as the ripples in the stopband. In the ideal case, we would want there to be no ripples in the passband

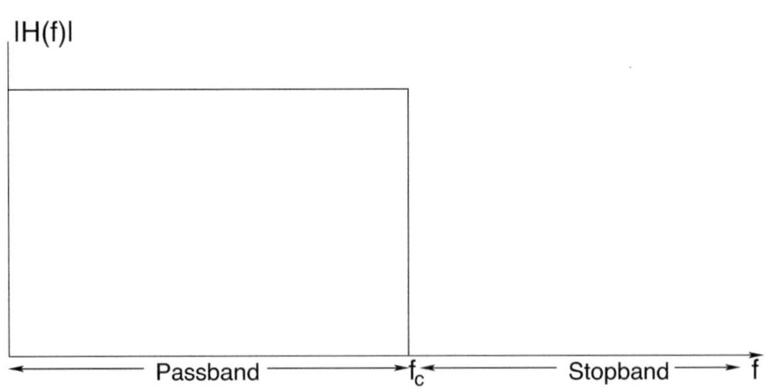

Fig. 3.11 An ideal low pass filter

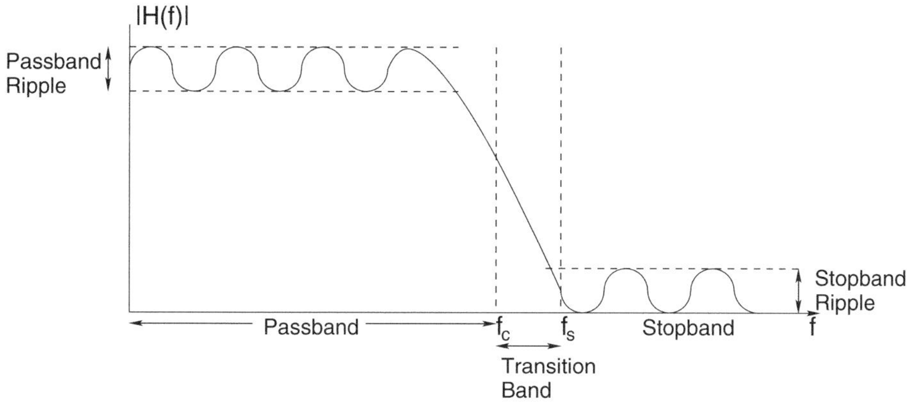

Fig. 3.12 Some of the parameters used in the design of filters

or stopband, and we would want the width of the transition band to be zero. In practice, different designs will achieve these ideal requirements to different degrees.

For example, the family of Butterworth filters, developed by Butterworth et al. [8], has the flattest passband of any filter family. The magnitude of the nth order low-pass filter is given by

$$|H(\omega)| = \frac{G}{\sqrt{1 + (\omega/\omega_c)^n}}$$

where G is the gain of the filter and ω_c is the cutoff frequency in radians per second. The transfer function can be written as

$$H(s) = G \prod \frac{1}{(s - s_k)}$$

where

$$s_k = \omega_c e^{j\frac{(2k+n-1)\pi}{2n}}, \quad k = 1, 2, \ldots, n$$

are the poles of the filter. Notice that the filter's poles are on the circle of radius ω_c in the left half s-plane. For example, for a fourth-order Butterworth filter, the poles, as seen in Fig. 3.13, are at angles of $5\pi/8$, $7\pi/8$, $9\pi/8$, and $11\pi/8$. All poles are in the left half plane, thus making the filter stable.

The attractive aspect of a Butterworth filter is its relatively flat passband. However, while it has a relatively flat passband, as shown in Fig. 3.14, it has a very large transition band. Figure 3.14 shows a plot of the magnitude and phase of an eighth-order Butterworth filter.

We can reduce the size of the transition band by relaxing the emphasis on a flat passband and stopband. Chebyshev filters have a much sharper roll-off than the Butterworth filters. However, they have ripples in either the passband or the stopband. There are two types of Chebyshev filters, Type I and Type II, with Type I being the more widely used. The Type I filter has ripples in the passband, and the Type II filter has ripples in the stopband. The

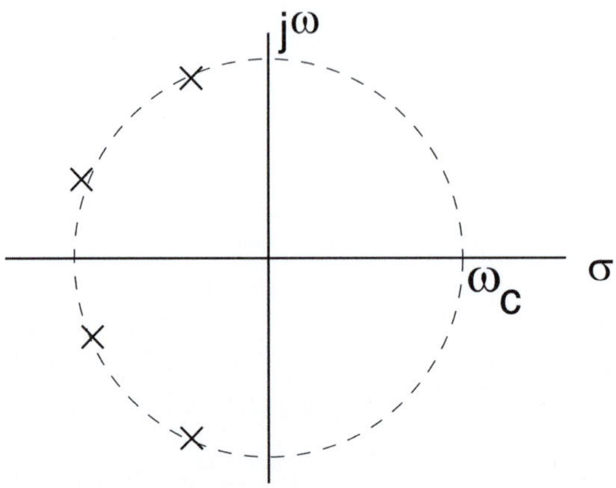

Fig. 3.13 Poles of a fourth-order Butterworth filter

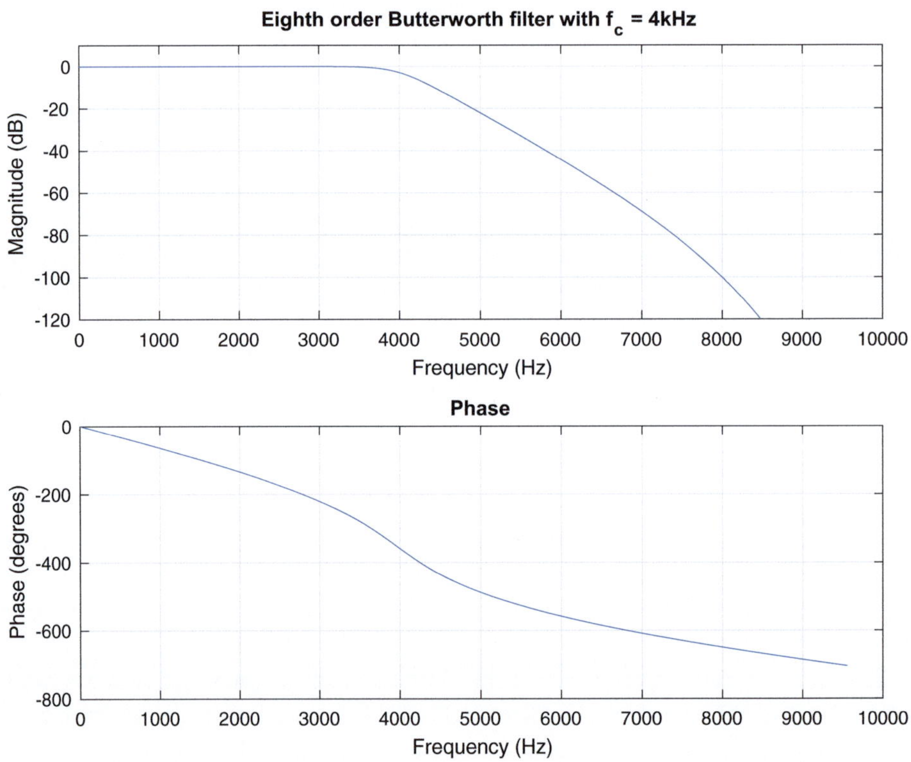

Fig. 3.14 Magnitude and phase of an eighth-order Butterworth filter with a cutoff frequency of 4 kHz

transfer functions of Chebyshev filters have a somewhat more complicated appearance than those of Butterworth filters. However, as there are routines available in most environments for implementing these filters in the digital domain, this is not a major hurdle. The transfer function of the nth order Type I Chebyshev filter is given by

$$|H(\omega)| = \frac{1}{\sqrt{1 + \epsilon^2 T_n(\omega/\omega_c)}}$$

where ϵ is the size of the passband ripple and $T_n(x)$ is the Chebyshev polynomial defined recursively as follows:

$$T_0(x) = 1$$
$$T_1(x) = x$$
$$T_n(x) = 2xT_{n-1}(x) - T_{n-2}(x)$$

The transfer function of the nth order Type II Chebyshev filter is given by

$$|H(\omega)| = \sqrt{\frac{\epsilon^2 T_n^2(\omega_c/\omega)}{1 + \epsilon^2 T_n^2(\omega_c/\omega)}}$$

These are rather messy-looking expressions. Luckily, for the most part, we don't have to deal with them directly. And when we need them, they are readily available. In Fig. 3.15 we show the magnitude and phase of an eighth-order Chebyshev Type I filter.

Comparing this with the eighth-order Butterworth filter, we can see that the transition from the passband is significantly sharper, and the transition band itself is significantly smaller than that of the Butterworth filter.

The transition band is even smaller in the Elliptic filter, so named because the poles lie on an ellipse. The tradeoff is that, unlike the Chebyshev filters, Elliptic filters have both passband and stopband ripple. Consider the magnitude of an eighth-order elliptical filter in Fig. 3.16. The transition between the passband and the stopband is almost instantaneous. However, there is a significant ripple in both the passband and the stopband.

Once we have the transfer function for the analog filter, we can convert it into a transfer function for a digital filter using one of several methods. There are different methods that work best for FIR and IIR filters. Let's look at the methods for each type of filter in turn.

3.3.1 Design of FIR Filters

For the FIR filters, we look at two common methods—the impulse response truncation method and the Parks-McClellan algorithm.

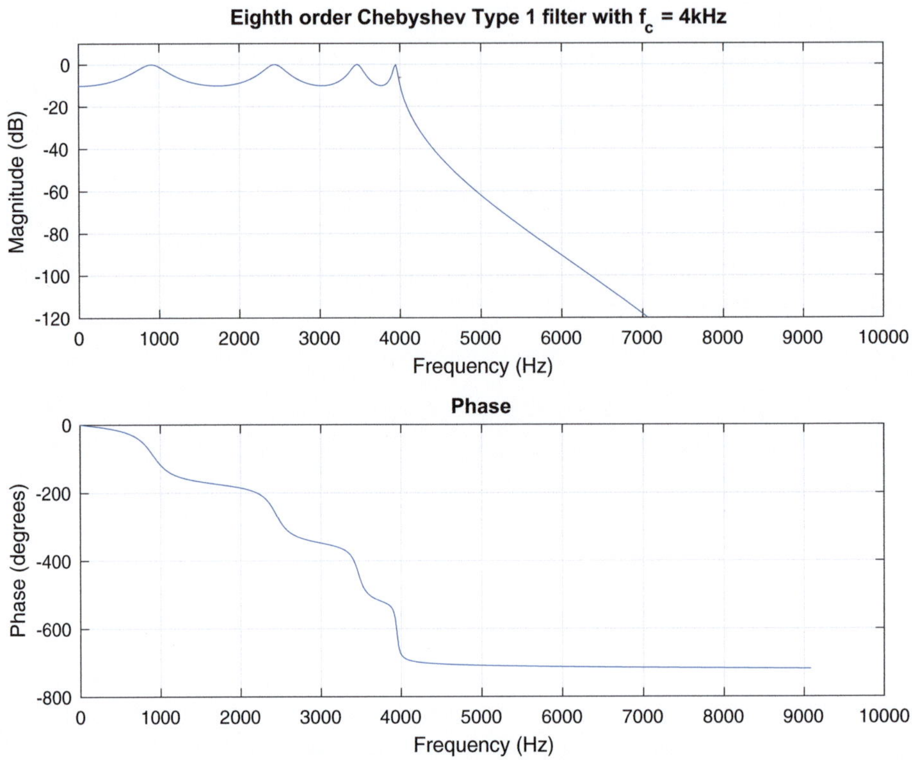

Fig. 3.15 Magnitude and phase of an eighth order Chebyshev Type I filter with a cutoff frequency of 4 kHz

Impulse Response Truncation Method

Recall that for the FIR filter

$$h[n] = \begin{cases} a_n \; n = 0, 1, \ldots, M \\ 0 \quad \text{otherwise} \end{cases}$$

In other words, the filter coefficients can be found directly from the desired impulse response. We can find an analog impulse response we like and simply set the filter coefficients to the corresponding sampled impulse response values. This is definitely a valid approach, though there are a few issues with it. Assume we wanted to design a low-pass filter. The ideal low-pass filter has a magnitude of one in the passband and a magnitude of zero in the stop band. The corresponding impulse response is given by

$$h[n] = \frac{\sin(n\omega_c)}{n\omega_c}$$

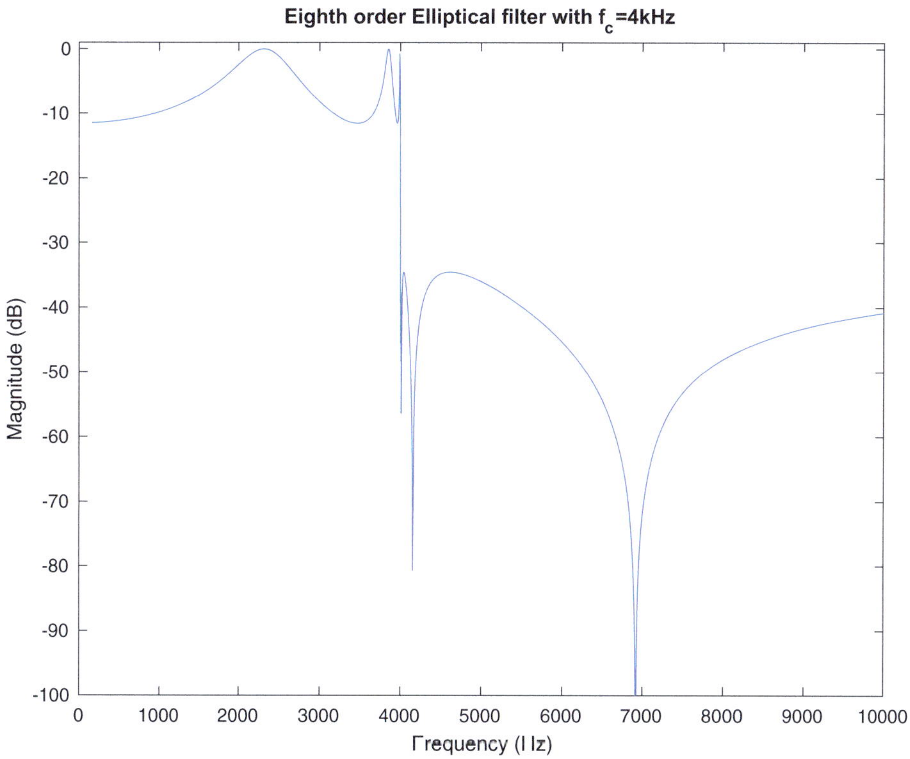

Fig. 3.16 Magnitude and phase of an eighth order Elliptic filter with a cutoff frequency of 4 kHz

shown in Fig. 3.17. Just setting $a_n = h[n]$ will not work for several reasons. The impulse response is nonzero from negative infinity to infinity. Therefore, if we want to build a finite impulse response filter, we will need to truncate this response. Furthermore, the impulse response is nonzero for $n < 0$, indicating that the system is not causal. So what happens if we simply truncate the impulse response and set all values of $h[n]$ for $n < 0$ to zero? The resulting transfer function magnitude is shown in Fig. 3.18.

This is not particularly good. We can do a bit better if we introduce a delay shifting the impulse response to the right and then zeroing out all the coefficients for $n < 0$ as shown in Fig. 3.19.

The corresponding transfer function is shown in Fig. 3.20. This is slightly better, though not by much.

The transfer function can be improved further if we are willing to tolerate more delay. Shifting the impulse response further, we get the response shown in Fig. 3.21. We trade off delay for better frequency characteristics. This method is called the *impulse response truncation (IRT)* method. In our example, we have simply truncated the impulse response.

Fig. 3.17 The impulse response corresponding to an ideal low pass filter

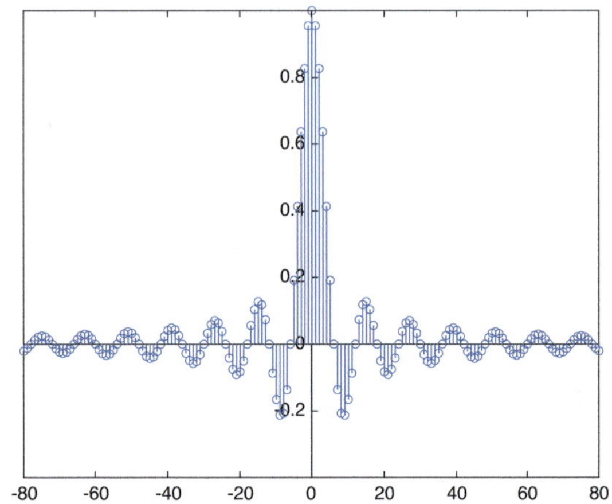

Fig. 3.18 The transfer function if we set $h[n] = 0$ for $n < 0$

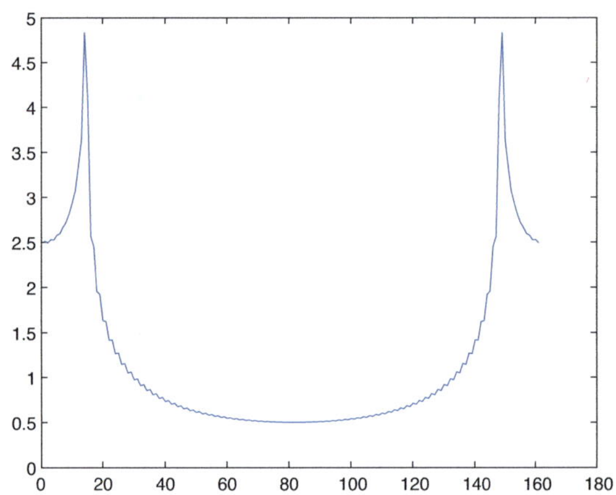

Rather than just a sharp truncation, we can also use one of the many windows to do a more graceful truncation, which leads to a reduction in the amount of ringing.

Parks-McClellan Algorithm

Another method for designing FIR filters is the one developed by James McClellan in Thomas Parks' graduate-level "Signal Theory" class at Rice University [9]. The algorithm is known as the Parks-McClellan algorithm [10]. Let's take a broad overview of how the algorithm works. We recognize that ideal filter characteristics are generally impossible to implement in practical filters. What we try to do instead is minimize the difference between the ideal characteristics and our filter. Let's suppose the desired transfer function is given by $H_d(\omega)$ and the transfer function of the filter we design is given by $H(\omega)$. Then, what we want to do is minimize the difference between $H(\omega)$ and $H_d(\omega)$. Generally, certain frequency ranges

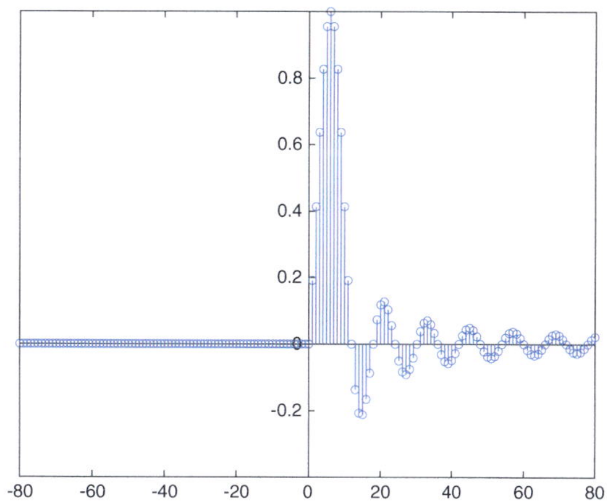

Fig. 3.19 A shifted impulse response with $h[n] = 0$ for $n < 0$

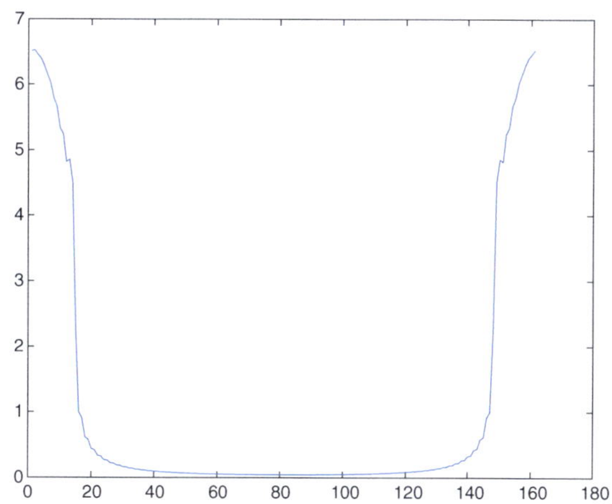

Fig. 3.20 The transfer function if we shift the impulse response by 6 samples and set $h[n] = 0$ for $n < 0$

are more important for our application, making it important that the difference between the desired and actual response be minimized more strictly in those regions. This variation of importance as a function of frequency can be represented by a weighting function $W(\omega)$. We can define an error metric

$$E(\omega) = W(\omega)\,|H_d(\omega) - H(\omega)|$$

and find the impulse response that minimizes this metric. Parks and McClellan focused on linear-phase Chebyshev filters. As we saw earlier, these filters have a sharper rolloff than most other filters but exhibit a ripple in the passband. FIR filters have a linear phase in four conditions:

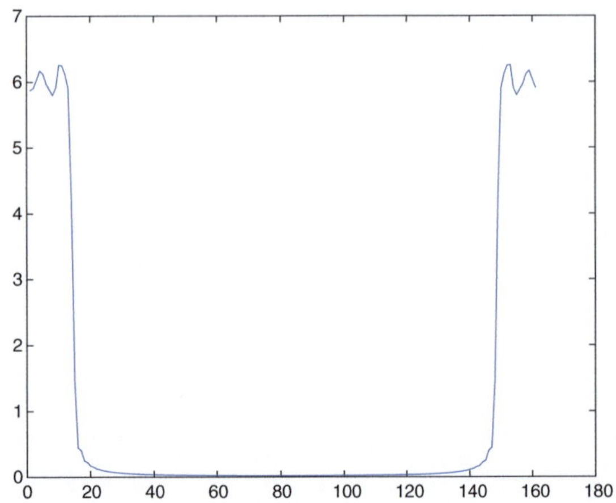

Fig. 3.21 The transfer function if we shift the impulse response by 24 samples and set $h[n] = 0$ for $n < 0$

$$\begin{aligned}
&\text{Type 1} \quad M \text{ odd} \quad h[n] = h[M - n] \\
&\text{Type II} \quad M \text{ even} \quad h[n] = h[M - n] \\
&\text{Type III} \quad M \text{ odd} \quad h[n] = -h[M - n] \\
&\text{Type IV} \quad M \text{ even} \quad h[n] = -h[M - n]
\end{aligned}$$

The fact that we are dealing with linear phase filters means that the filter transfer response is given by

$$H(\omega) = G(\omega)e^{-jK\omega}$$

where $G(\omega)$ is real. The fact that we are dealing with Chebyshev filters means we can utilize an algorithm developed for use with Chebyshev polynomials by Remez [11], appropriately known as the *Remez algorithm*, for minimizing the error metric. The algorithm iteratively tries out different impulse response values based on the Chebyshev polynomials until no further improvement is available.

In MATLAB, the Parks-McClellan algorithm is implemented in the function *firpm*. In its simplest form, this function takes three inputs: the order of the filter, a vector of the extremal frequencies normalized by half the sampling frequency, and the ideal shape. By extremal frequencies, we mean the ends of the passbands and stopbands. Let's look at a couple of examples.

Suppose we have a signal sampled at 1000 samples per second, and we want to implement a low-pass filter of order 30, which will pass all frequencies up to 200 Hz and stop all frequencies above 300 Hz. In other words, the passband spans from 0 to 200 Hz, the transition band ranges from 200 Hz to 300 Hz, and the stopband extends from 300 Hz to 500 Hz. Dividing by half the sampling frequency, or 500, we get the normalized extremal frequencies 0/500, 200/500, 300/500, and 500/500, or 0, 0.4, 0.6, and 1. To construct the idealized

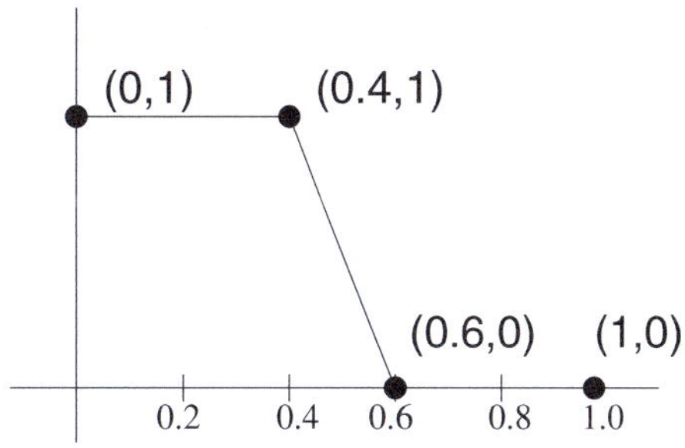

Fig. 3.22 Ideal transfer function characteristics

spectrum, we connect the point (0, 1) to (0.4, 1) and the point (0.6, 0) to (1, 0) to get the spectrum shown in Fig. 3.22.

If we label the extremal points as (x_1, y_1), (x_2, y_2), (x_3, y_3) and (x_4, y_4), the MATLAB command is written as $firpm(order, [x_1 x_2 x_3 x_4], [y_1 y_2 y_3 y_4])$. For this particular example, this becomes

```
h = firpm(30,[0 0.4 0.6 1],[1 1 0 0]);
```

The resulting impulse response and transfer function magnitude are shown in Fig. 3.23.

To make a bandpass filter with the pass band between 200 and 300 Hz, with the same sampling rate and the same width of transition bands on both sides of the passband, the command

```
h = firpm(30,[0 .2 .4 .6 .8 1], [0 0 1 1 0 0]);
```

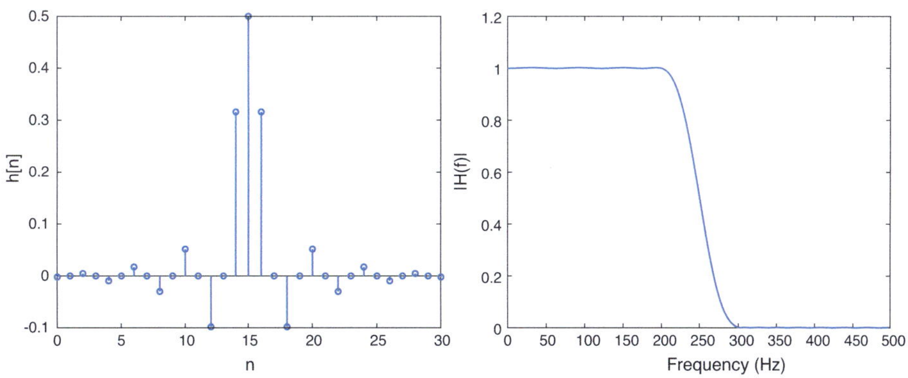

Fig. 3.23 Impulse response and transfer function of an FIR low-pass filter designed using the Parks-McClellan algorithm

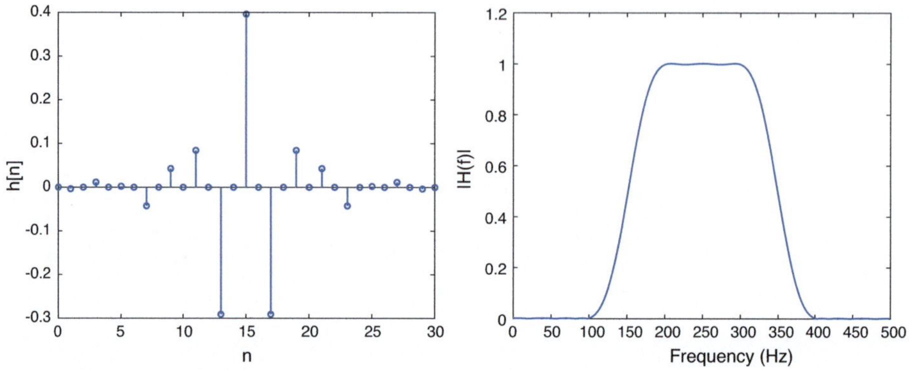

Fig. 3.24 Impulse response and transfer function of an FIR band pass filter designed using the Parks-McClellan algorithm

will produce the filter shown in Fig. 3.24. These look significantly better than the filters we obtained using the impulse response truncation method.

3.3.2 Design of IIR Filters

One way of designing IIR filters is the way we have already used: placing zeros at the frequencies we wish to block and placing poles to either cancel out the effect of the zeros or to enhance the frequencies we are interested in. Often, you have a very simple task at hand: you want to remove line interference or the very low-frequency effect due to baseline wander. In such cases, placing poles and zeros according to the specific problem at hand can be quite effective. However, if you need a more sophisticated approach, you can take one of the well-known analog filters with known poles and zeros and transfer the poles and zeros from the s domain to the Z domain.

Impulse Invariance Method

The impulse invariance method is in a sense the counterpart of the Impulse Response Truncation method for FIR filters. We start with a partial fraction expansion of the transfer function

$$H(s) = \sum_{k=1}^{n} \frac{A_k}{s - s_k}$$

From this we can find the impulse response as

$$h(t) = \sum_{k=1}^{n} A_k e^{s_k t} u(t)$$

Sampling the impulse response every T_s seconds and normalizing $T_s = 1$ we get

$$h[n] = \sum_{k=1}^{n} A_k e^{s_k n} u[n]$$

From which we can get the Z transform as

$$H(z) = \sum_{k=1}^{n} \frac{A_k}{1 - e^{s_k} z^{-1}}$$

Expanding this out, we obtain the transfer function in terms of z^{-1} from which we can get the difference equation.

Because we are sampling in the time domain, this means that we will get replication in the frequency domain. Because no real system is perfectly band-limited, this means that we will get aliasing distortion in the response. We can minimize this by ensuring that the sampling frequency is high enough and the stopband magnitude is low enough at the sampling frequency to cause minimal trouble.

Example Let's use the impulse invariance method to obtain the digital equivalent of a second-order Butterworth filter, that is $n = 2$. This is a pretty low order, and the results are not going to be particularly pleasing, but we can see the process. Let's pick a normalized cutoff frequency of 0.5, then using the fact that the poles of the Butterworth filter are given by

$$s_k = \omega_c e^{j\frac{(2k+n-1)\pi}{2n}}, \quad k = 1, 2, \ldots, n$$

we have

$$s_1 = 0.5e^{j\frac{3}{4}\pi}$$
$$s_2 = 0.5e^{j\frac{5}{4}\pi}$$

or

$$s_1 = -\frac{1}{2\sqrt{2}} + j\frac{1}{2\sqrt{2}}$$
$$s_2 = -\frac{1}{2\sqrt{2}} - j\frac{1}{2\sqrt{2}}$$

from which we can obtain the poles of the digital filter as

$$z_1 = e^{s_1} = e^{-\frac{1}{2\sqrt{2}}} e^{j\frac{1}{2\sqrt{2}}}$$
$$z_2 = e^{s_2} = e^{-\frac{1}{2\sqrt{2}}} e^{-j\frac{1}{2\sqrt{2}}}.$$

Bilinear Z Transform

There is a somewhat more conceptually complicated method that avoids aliasing. This is when we use the bilinear Z transform. In this transform, we again begin with analog filters. Instead of sampling the impulse response we use the transform

$$s = \left(\frac{2}{T_s}\right)\frac{1 - z^{-1}}{1 + z^{-1}}$$

to move between the Z domain and the s domain.

If we work out all the math we find that a point $s = \sigma + j\omega$ in the s-plane gets mapped into the point $re^{j\theta}$ in the Z plane where

$$r = \frac{\sqrt{\left(1 + \frac{\sigma T_s}{2}\right)^2 + \left(\frac{\omega T_s}{2}\right)^2}}{\sqrt{\left(1 - \frac{\sigma T_s}{2}\right)^2 + \left(\frac{\omega T_s}{2}\right)^2}}$$

and

$$\theta = \arctan\left[\frac{\left(\frac{\omega T_s}{2}\right)}{\left(1 + \frac{\sigma T_s}{2}\right)}\right] + \arctan\left[\frac{\left(\frac{\omega T_s}{2}\right)}{\left(1 - \frac{\sigma T_s}{2}\right)}\right]$$

To understand how this mapping works let's simplify the transformations by setting $T_s = 2$. With this value of T_s the mapping is simplified to:

$$r = \frac{\sqrt{(1 + \sigma)^2 + \omega^2}}{\sqrt{(1 - \sigma)^2 + \omega^2}}$$

$$\theta = \arctan\left[\frac{\omega}{1 + \sigma}\right] + \arctan\left[\frac{\omega}{1 - \sigma}\right].$$

Recall that for a stable filter, the poles must be located in the left half-plane. The left half of the s-plane consists of all points in the plane for which $\sigma < 0$. If we look at the expression for r, we can see that the numerator and denominator are very similar except that the numerator has the term $(1 + \sigma)^2$ while the denominator has the term $(1 - \sigma)^2$. This means that when σ is negative, the denominator will be larger than the numerator, and hence r will have a magnitude less than one. There is no particular effect of σ being negative on the value of θ. This means that points in the left half s-plane get mapped to points inside the unit circle in the Z domain. If we examine the mapping for the case where σ is greater than zero, that is, the right half of the s-plane, we can see that all points in the right half of the s-plane map to points outside the unit circle. This means that under this mapping, stable analog filters map to stable digital filters. The $j\omega$ axis is where σ is zero. As we can see from the mapping, when $\sigma = 0$,

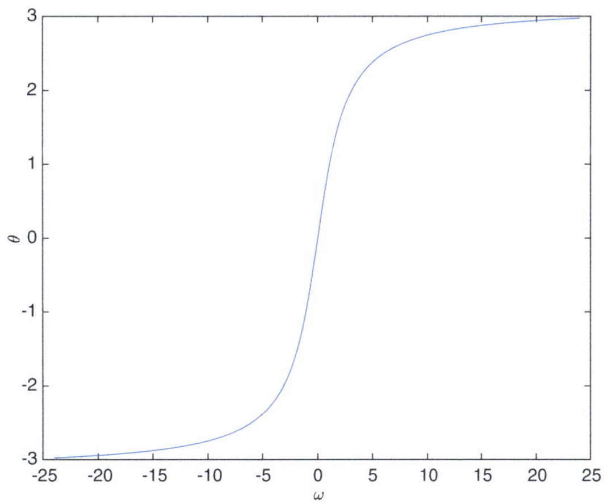

Fig. 3.25 Plot of the analog frequency versus the digital frequency after the bilinear Z transform. Notice that the mapping from ω to θ is one-to-one

$$r = \frac{\sqrt{(1+\omega^2}}{\sqrt{1+\omega^2}} = 1$$

and the $j\omega$ axis is mapped into the unit circle. The advantage of this mapping from the $j\omega$ axis to the unit circle is that it is one-to-one. We can see this if we set σ to zero and plot ω vs θ as shown in Fig. 3.25. This means we avoid the problem of aliasing.

MATLAB In practice, we can utilize one of the various tools available in MATLAB for designing digital filters using the bilinear transform. Suppose we wanted to use a low-pass Butterworth filter. If we know the order of the filter we want to use and the cutoff frequency f_c we can use the function *butter(N,Fc)* where N is the order of the filter, and Fc is the normalized cutoff frequency

$$Fc = \frac{f_c}{(f_s/2)}$$

where f_s is the sampling frequency. The function returns the transfer function of the filter in terms of the coefficients of the numerator and denominator polynomials. Suppose the transfer function of the filter is

$$H(z) = \frac{b_1 + b_2 z^{-1} + \cdots + b_{n+1} z^{-n}}{a_1 + a_2 z^{-1} + \cdots + a_{n+1} z^{-n}}$$

where a_i and b_i are components of the vectors A and B, then

$$[B, A] = butter(N, Fc)$$

returns the coefficients of the filter. The program translates the requirements to first obtain an analog lowpass Butterworth filter. It then uses the bilinear transform described above to obtain the digital filter.

If we are unsure of the order of the filter and the cutoff frequency but we know that we want to definitely pass the frequencies between 0 and f_p and stop the frequencies greater than f_s and we can further specify the maximum passband ripple Rp and the minimum amount of attenuation in the stopband Rs then we can obtain the order and cutoff frequencies as

$$[N Fc] = buttord(Fp, Fs, Rp, Rs)$$

The values of N and Fc can then be used in the function *butter* to obtain the discrete-time filter. Once we have the filter coefficients, we can implement the filter as a difference equation. We can also use the *filter* command in MATLAB to implement the filter.

Example: Let's suppose we have a sum of three sinusoids at 10 Hz, 30 Hz, and 60 Hz, sampled at 250 samples per second. We want to construct a Butterworth low-pass filter, which will let through the first two sinusoids and block the last one. The following MATLAB code outlines the steps described above to construct a filter and provides an implementation of the filter to achieve the task of allowing the first two sinusoids through and blocking the last one.

```
clear all
fs = 250; % sampling rate in samples persecond
tD = 1; % duration of the signal
N = tD*fs; % Total number of samples
t = [0:1/N:tD]; % sampling times

fpass = 30;  % This is the breakpoint for the passband in Hz

fstop = 50;   % This is the breakpoint for the stopband in Hz

Fp = fpass/(fs/2);
Fs = fstop/(fs/2);
Rp =1;
Rs = 30;
[Order, Fc] = buttord(Fp, Fs, Rp, Rs); % Order of the filter for the
                                        % specified parameters.
[B, A] = butter(Order,Fc);  % numerator and denominator coefficients
                            % of the transfer function.

% Use this filter to process a signal composed of three sinusoids with
% frequencies f1, f2, and f3

% Generate the input signal
f1 = 10; % frequency in Hz of the first sinusoid
f2 = 30; % frequency in Hz of the second sinusoid
f3 =60;  % frequency in Hz of the third sinusoid
x = cos(2*pi*f1*t) + 1.5*cos(2*pi*f2*t) + cos(2*pi*f3*t); % generate
the input
M=N; %FFT size
```

```
y = fft(x,M);
% To obtain the frequency axis recall that the DFT samples the
% DTFT from 0 to 2pi in increments of 1/M.  Also recall that 2pi
% corresponds to the sampling frequency.  Therefore, each increment
% in frequency is fs/M, and the highest frequency is the sampling
% frequency fs
freqAxis=[0:fs/M:(N - M/N)/2];  % When we pick M to be different than
                                % N the number of points in this array
                                % may be different than the number of
                                % points in the t array; adjust accordingly
figure
subplot(4,1,1)
plot(t,x);
title('Original Signal');
subplot(4,1,2)
plot(freqAxis,abs(y(1:M/2)))
title('Magnitude of the DFT of the original signal');

z = filter(B,A,x);          % filter the input.
w = fft(z,N);  % Find the DFT of the output
subplot(4,1,3)
plot(t,z)
title('Filtered Signal')
subplot(4,1,4)
plot(freqAxis,abs(w(1:M/2)))
title('Magnitude of the DFT of the filtered signal')
```

The output of this program is shown in Fig. 3.26.

The procedure described above can also be used to design Chebyshev and Elliptical filters. The counterparts to *butter* are *cheby*1 and *cheby*2 for Chebyshev Type I and Type II filters, and *ellip* for Elliptical filters. The counterparts for *buttord* are *cheb*1*ord* and *cheb*2*ord* for Chebyshev Type I and Type II filters, and *ellipord* for Elliptical filters.

Finite Wordlength Effect

Digital filters are not just discrete-time filters; they are also *digital* filters. This means that the coefficients cannot take any possible value. To be able to handle any possible value, you would need to have an infinite number of bits. In practice, the number of bits, or the word length, is finite. This means that the coefficients have to be quantized. This quantization can result in unwanted effects. Consider a simple filter designed to severely attenuate the DC component of a signal without affecting the other frequency components. To achieve this, we place a zero at $z = 0.98$ and a pole at $z = 0.97$. The transfer function of the filter is

$$H(z) = \frac{z - 0.98}{z - 0.97}$$

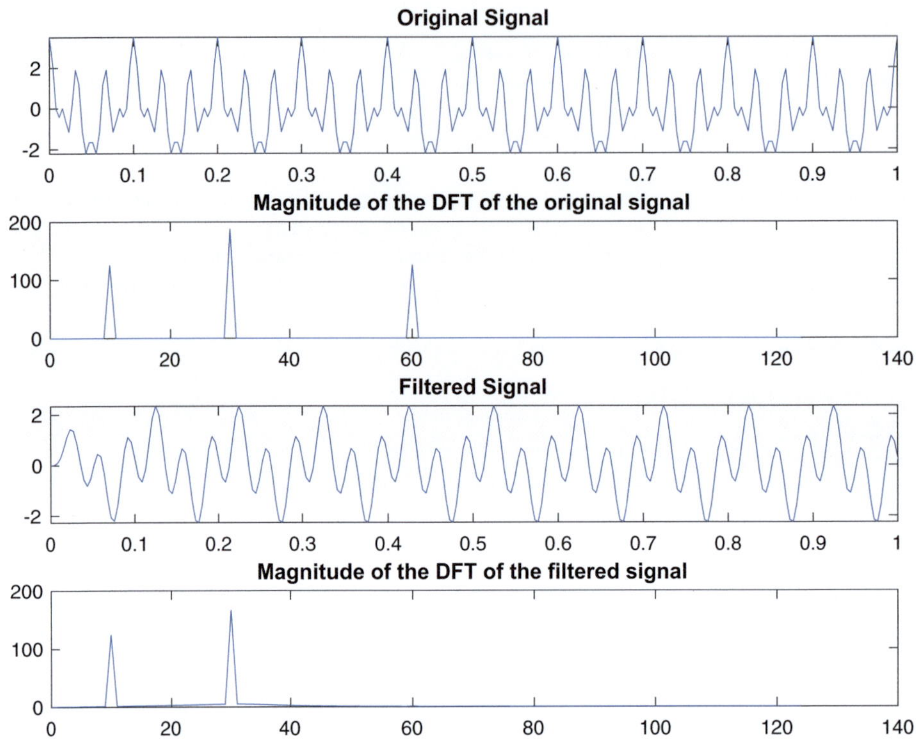

Fig. 3.26 Output of the MATLAB program

and the corresponding difference equation is

$$y[n] = x[n] - 0.98x[n-1] + 0.97y[n-1]$$

As we can see from Fig. 3.27 this simple filter does a decent job of attenuating the very low frequencies without attenuating the higher frequencies.

Suppose we want to implement this filter using a system that only allows us six bits to encode the filter coefficients. If we assume for simplicity that the magnitude of the coefficients falls between 0 and 1, we have to partition the unit interval into 64 regions of size 1/64 and assign all numbers that fall into each partition the same value. In the case of our simple filter, both the values of 0.97 and 0.98 fall into the same interval, resulting in the pole and zero values canceling each other out. The filter now does nothing! This effect is a result of the finite wordlength, or the finite number of bits, used to represent numbers in digital filters. In this particular case, the result was relatively benign. However, we can

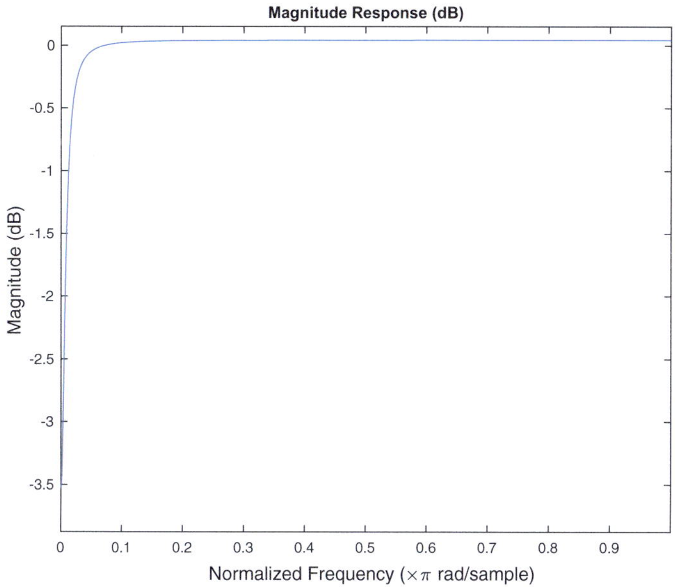

Fig. 3.27 Magnitude of a simple filter with a zero at 0.98 and a pole at 0.97

end up with situations where, because of the finite wordlength, the placement of poles and zeros is such that the filter goes unstable. This is particularly something to be aware of when implementing digital filters in embedded systems with limitations on the wordlength.

3.4 Summary

In this chapter, we have provided a somewhat cursory review of digital filters. Most software applications, such as MATLAB or Python, will have routines for designing and implementing digital filters. Given this, was even a cursory review of digital filters necessary? Biomedical signal processing is a diverse field with numerous applications. Therefore, our assumption is that it is better to have more information than less. The basic takeaway from this chapter is that if we want to achieve a simple task, such as eliminating line frequencies or reducing the effect of baseline wander, it is straightforward to design a simple filter by placing poles and zeros at the appropriate locations. For anything more complex, it is a good idea to use the various tools available to design the filters.

4 Features

Many biomedical signal processing applications involve distinguishing between signals generated under different conditions. We might want to differentiate between the EEG signals from people when they are having a seizure and the signals when they are not having a seizure. Or, perhaps, we want to differentiate between EKG signals indicating tachycardia from normal EKG signals. Given a set of measurements, we want to select the process that generated those measurements from a set of candidate processes. The selection is based on the features of the processes that distinguish them from one another. If these features are measurable, we can construct a classifier that categorizes observations into distinct classes. We will talk about classifiers and their properties in a later chapter. For now, let's focus on the idea of features.

Suppose we were looking to select people for training as football players based on their responses to a questionnaire about their physical characteristics—their physical features. Perhaps we would start with height and weight. We can plot each individual's height and weight in two dimensions and then partition the space with a straight line to distinguish between acceptable and unacceptable values. This would work to a degree. You would reject people who were too skinny to withstand a blow, too overweight to be able to move fast, or too short for the positions you were thinking of. Of course, you could have somebody skinny, or overweight, or short who would make a great football player—but a classifier does not have to be perfect. On the other hand, you might very likely not reject me—6' 1" and 195 pounds. But just as clearly, I would not be a good candidate for playing football. Perhaps you said, let's add strength as another feature, measured by the amount of weight you can lift. Now we have three dimensions—height, weight, and strength- which define a point in three-dimensional space. We can now partition this three-dimensional space with a plane and build a classifier that would immediately eliminate people like me. But suppose—and this is highly theoretical—I got a gym membership and started pumping iron, increasing

K. Sayood, *Biomedical Signal Processing*, Synthesis Lectures on Biomedical Engineering,
https://doi.org/10.1007/978-3-032-08855-0_4

my strength. Just to mess up the classifier. What now? Well, we could let people like me proceed on their suicidal quest, or we could add another dimension to our classifier. Say of age. That would definitely and finally take people like me out of contention.

What we have been doing is selecting features appropriate to the task at hand. We start with a goal: dividing a set of objects (people) into classes (football players and non-football players). To do this classification, we need to have measurable quantities. These quantities we call features. In many classification problems, as in this case, there are features specific to the application at hand that are best utilized. For example, the distance between peaks may be a suitable feature to use in an EKG application for detecting tachycardia. The energy in different frequency bands may be a useful feature to utilize in an EEG application for detecting seizures. We will look at these features when we discuss those particular applications. Here, we will examine features that are not necessarily application-dependent and can be utilized for a wide range of applications. Application-agnostic features, for the most part, can be divided into two broad classes: statistical features and matrix-based features. We look at some examples of features from each of these classes in this chapter.

4.1 Statistical Features

The most easily extracted and, therefore, the most widely used application-agnostic features in classification applications are statistical features such as mean, median, variance, and correlation. Let's briefly look at each of these. As a motivating toy example, we will use six examples of a male speaker saying "hello" and six examples of a female speaker saying the same. The goal is to build a classifier that can classify a speaker as male or female based on their utterance of "hello." The waveforms are shown in Fig. 4.1.

Mean If we treat the experiment generating the data being considered as a random process, then the mean or the average value of the samples can be defined in terms of the probability density of the random variable as the expected value of the random variable. In practice, the mean is computed as

$$\hat{\mu} = \frac{1}{N} \sum_{i=1}^{N} x_i$$

where x_i are data points and N is the number of data points. The average values of the speech signals shown in Fig. 4.1 are shown in Table 4.1. While the average values of various samples vary slightly, effectively, there is no difference between the male and female sets. The mean value would not be a good feature to use in our application.

Median The median of a sequence of data values is that value for which half of the data is above that value and half is below it. For an odd number of data values, this is usually the middle value in the ordered array of values. If the number of data values is even, we typically calculate the median by taking the average of the two middle values. In situations where a few extreme values can substantially skew the mean, the median is a much better indicator

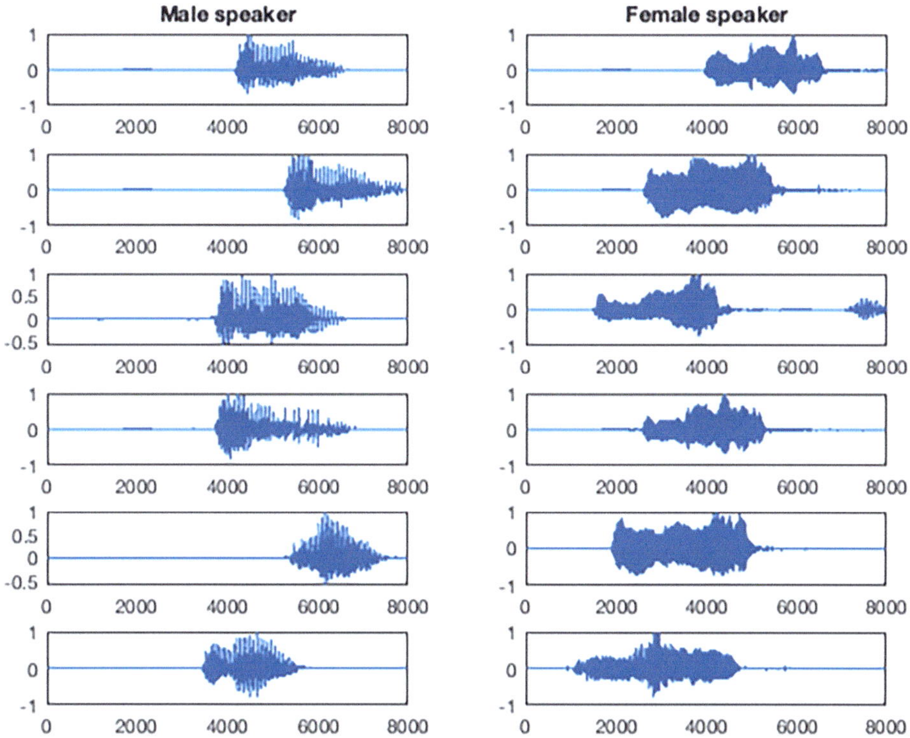

Fig. 4.1 Six examples of a male speaker saying hello and a female speaker saying hello

Sample	Male speaker	Female speaker
1	0.000306	0.000186
2	0.000533	0.000051
3	0.000400	0.000303
4	0.001017	0.000009
5	0.000392	0.000453
6	0.000463	0.000495

of the relative magnitude of the data. Suppose you have a bar where techies gather. Ten engineers are gathered in a bar one night, and their yearly incomes, in thousands of dollars, are 80, 87, 86, 84, 92, 90, 92, 93, 95, and 101. If we compute their average yearly income, it comes to 90K. The median in this case is 91K. Suppose now the oligarch Bill Gates walks in. The average annual income for the folks in the bar suddenly shot up to about 455 million dollars! This is not at all an accurate reflection of the earnings of the people in the bar. The median, however, is now 92K. The median gives us a better idea of the yearly income of the folks in the bar. In the case of our toy example, the median will likely be around zero for

Table 4.1 Variance of speech samples from male and female speakers

Sample	Male speaker	Female speaker
1	0.012327	0.019747
2	0.015154	0.041418
3	0.018036	0.021844
4	0.016961	0.021072
5	0.009860	0.039337
6	0.013177	0.022885

both male and female speakers. The median would not be a good feature for classifying the speech as emanating from a male or female speaker.

Variance The variance is a measure of the variability of data values around the mean. Because data values that are more variable usually also contain more information, we may sometimes use the variance as a proxy for the "informativeness" of a particular kind of data. The estimate of the variance is usually computed as

$$\hat{\sigma}^2 = \frac{1}{N-1}\sum_{i=1}^{N}(x_i - \mu)^2$$

The division by $N - 1$ rather than N may not make a significant difference when N is large, but it can be shown that using $N - 1$ instead of N yields an *unbiased estimator*. An unbiased estimator is an estimator whose expected value is the same as the expected value of the parameter being estimated.

If we examine the variances for our toy example, we obtain the results shown in Table 4.1. The variances for the speech samples from the female speaker are generally higher than the variances of the speech samples from the male speakers. This particular feature appears to be helpful in differentiating between male and female speakers, at least in this toy example.

Correlation Sometimes, in a classification problem, we have representative signals available for the different classes. One way to classify a new signal is to determine which of the representative signals it is most closely related to. A simple measure of similarity is the correlation between the signals

$$\hat{\rho}_{XY} = \frac{\sum_{i=1}^{N}(x_i - \hat{\mu}_x)(y_i - \hat{\mu}_y)}{(N-1)\hat{\sigma}_x\hat{\sigma}_y}$$

Autocorrelation Autocorrelation is the correlation of the signal with itself, delayed by a certain amount

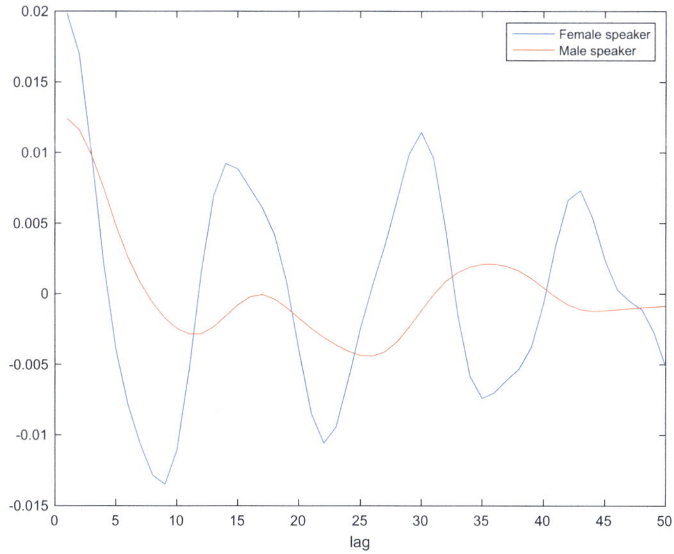

Fig. 4.2 Autocorrelation coefficients for one male and one female speaker saying "hello"

$$R_{xx}(k) = \frac{1}{N}\sum_{i=1}^{N} x_i x_{i-k}$$

The autocorrelation function is useful in several ways, as we shall see below. At a basic level, the autocorrelation function reflects the frequency structure of the data. If the signal changes faster, there will be less correlation between samples that are further apart than if the signal changes more slowly. In other words, high-frequency content tends to cause the autocorrelation to decay quickly. Low-frequency content will promote a slower decay in the value of the autocorrelation function. If the data is periodic, this periodicity is also reflected in the autocorrelation values. The autocorrelation functions of the speech segments from a male speaker and a female speaker for our toy example are shown in Fig. 4.2. By counting the peaks, we can see that the autocorrelation function for the female speaker corresponds to a signal with a higher frequency. Therefore, the autocorrelation function, or perhaps the number of peaks in a fixed number of lags, could be a useful discriminating feature.

Model-based features Models, in particular linear models, provide a very efficient way of representing the structure of the data in a few values. Consider a set of data that can be modeled as the output of a linear filter described by the input-output relationship.

$$x[n] = \sum_{i=1}^{M} a_i x[n-i] + \epsilon_n$$

where $\{\epsilon_n\}$ is a white noise process. Given the data $\{x[n]\}$, we can find the $\{a_i\}$, the coefficients of the models, as follows.

Multiplying both sides of the equation by $x[n-k]$ and then taking the average over N values, we get

$$\begin{aligned}\frac{1}{N}\sum_{n=1}^{N} x[n]x[n-k] &= \frac{1}{N}\sum_{n=1}^{N}\left(\sum_{i=1}^{M} a_i x[n-i] + \epsilon_n\right) x[n-k] \\ &= \sum_{i=1}^{M}\left(\frac{1}{N}\sum_{n=1}^{N} a_i x[n-i]x[n-k]\right) + \frac{1}{N}\sum_{n=1}^{N}\epsilon_n x[n-k]\end{aligned}$$

The term on the left-hand side of this equation is the autocorrelation $R_{xx}(k)$. The second term on the right-hand side goes to zero because of the nature of the $\{\epsilon_n\}$ process. If we assume that the correlation between a sample and its neighbor k samples to the left is the same as its correlation with its neighbor k samples to the right, then we can write the first term on the right-hand side as

$$\begin{aligned}\sum_{i=1}^{M}\left(\frac{1}{N}\sum_{n=1}^{N} a_i x[n-i]x[n-k]\right) &= \sum_{i=1}^{M} a_i\left(\frac{1}{N}\sum_{n=1}^{N} x[n-i]x[n-k]\right) \\ &= \sum_{i=1}^{M} a_i R_{xx}(|i-k|)\end{aligned}$$

Putting it all together, we get

$$R_{xx}(k) = \sum_{i=1}^{M} a_i R_{xx}(|i-k|)$$

Writing this out for values of k from 1 to M we get the following M linear equations

$$\begin{aligned}R_{xx}(1) &= a_1 R_{xx}(0) + a_2 R_{xx}(1) + \ldots a_M R_{xx}(M-1) \\ R_{xx}(2) &= a_1 R_{xx}(1) + a_2 R_{xx}(0) + \ldots a_M R_{xx}(M-2) \\ R_{xx}(3) &= a_1 R_{xx}(2) + a_2 R_{xx}(1) + \ldots a_M R_{xx}(M-3) \\ \vdots &= \vdots \\ R_{xx}(M) &= a_1 R_{xx}(M-1) + a_2 R_{xx}(M-2) + \ldots a_M R_{xx}(0)\end{aligned}$$

We can solve these M equations for the M coefficient values $\{a_i\}$, and then use these M values as proxies for the process $\{x[n]\}$.

In Fig. 4.3, we have plotted the two coefficients from second-order models for our toy example. We can see that the coefficients for the female speaker congregate in the lower right-hand corner of the plot. We can easily partition this space as shown in Fig. 4.4. In

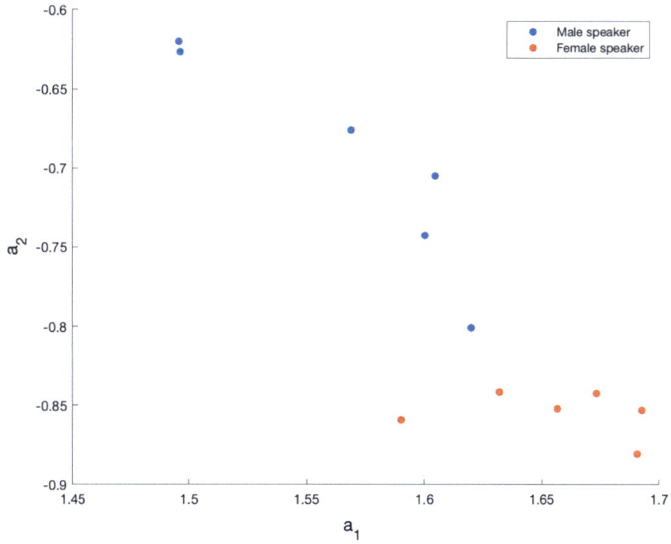

Fig. 4.3 Coefficients for second-order models for the audio sequences in the toy example

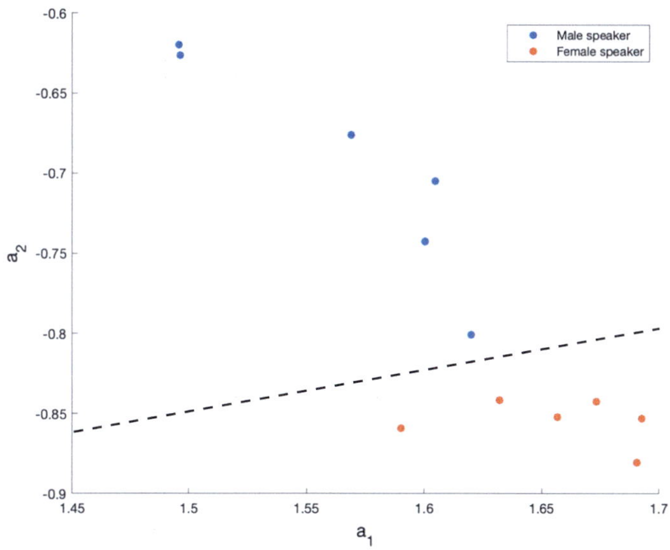

Fig. 4.4 Coefficients for second-order models for the audio sequences in the toy example

this particular example, using the coefficients of the second-order model should allow us to differentiate between the male speaker and the female speaker.

4.2 Matrix Interpretation

Many application-agnostic feature selection methods are based on a matrix interpretation of the data. The data collected is organized in matrix form, where the rows correspond to objects, for want of a better word, and the columns correspond to attributes. So, for an application that tries to classify students based on their test scores, as shown in Fig. 4.5, the rows would correspond to students, and the columns would correspond to exams.

In an application where we are trying to cluster genes that code for proteins with similar functions, the rows correspond to genes, and the columns may correspond to expression levels in different samples. In a search application, rows correspond to terms, and columns correspond to data. If we are trying to understand which patients are best served by a particular intervention, the rows would correspond to patients, and the columns might correspond to demographic data or test results. Due to the importance of matrices in various biomedical classification tasks, we will briefly review the aspects of the subject that are relevant to us.

4.2.1 A Brief Review of Matrices

Matrices are collections of mathematical objects organized in rows and columns. The mathematical objects can be mathematical expressions, real numbers, or complex numbers. As our examples above indicate, for us, matrices are generally collections of numbers organized into rows and columns. The size of a matrix is usually given as $n \times m$, where n is the number of rows and m is the number of columns. The element in the ith row and the jth column of a matrix $\mathbf{A}$ is denoted as $a_{i,j}$.

	Test 1	Test 2	Test 3	Test 4
Alice	9	4	10	8
Borges	10	1	10	2
Carol	10	5	10	10
David	3	4	7	8
Eva	2	5	4	10

$$\begin{bmatrix} 9 & 4 & 10 & 8 \\ 10 & 1 & 10 & 2 \\ 10 & 5 & 10 & 10 \\ 3 & 4 & 7 & 8 \\ 2 & 5 & 4 & 10 \end{bmatrix}$$

Fig. 4.5 Grades for a class of five students

Two matrices **A** and **B** can be multiplied to obtain a third matrix **C**

$$\mathbf{C} = \mathbf{A} \cdot \mathbf{B}$$

if the number of columns in **A** is equal to the number of rows of **B**. The number of rows of **C** is equal to the number of rows of **A**, and the number of columns of **C** is equal to the number of columns of **B**. The element $c_{i,j}$ is given by

$$c_{i,j} = \sum_k a_{i,k} b_{k,j}$$

Multiplication of matrices is not a commutative operation; that is,

$$\mathbf{A} \cdot \mathbf{B} \neq \mathbf{B} \cdot \mathbf{A}$$

The operation corresponding to division in this world is very different from scalar division. To describe the division of matrices, we first need to define the inverse of a matrix. The inverse of a matrix **A**, denoted by $\mathbf{A}^{-1}$, is the matrix which, when multiplied by **A**, gives us the identity matrix **I**.

$$\mathbf{A}^{-1}\mathbf{A} = \mathbf{A}\mathbf{A}^{-1} = \mathbf{I}$$

where the identity matrix **I** is a square matrix with ones on the main diagonal and zeros everywhere else. The inverse of a matrix **A** only exists if **A** is a square matrix and if the rows and columns of **A** are independent. By a square matrix, we mean that the number of rows in the matrix equals the number of columns. By independent, we mean that we cannot obtain any row or column as the linear combination of other rows and columns.

4.2.2 A Matrix is a Mapping

We can view a matrix as a mapping that provides a different representation of a set of values. If we organize the values in a vector **x** then the matrix A can map **x** into the vector of values **y** through multiplication

$$\mathbf{y} = A\mathbf{x}$$

Consider the matrix

$$A = \begin{bmatrix} \frac{1}{2} & \frac{1}{2} \\ \frac{1}{2} & -\frac{1}{2} \end{bmatrix}$$

Suppose we have a two-dimensional vector **x** with components x_1 and x_2. Then,

$$\begin{bmatrix} y_1 \\ y_2 \end{bmatrix} = \begin{bmatrix} \frac{1}{2} & \frac{1}{2} \\ \frac{1}{2} & -\frac{1}{2} \end{bmatrix} \begin{bmatrix} x_1 \\ x_2 \end{bmatrix}$$

is a mapping of **x** to the vector **y**. Examining the components of **y** we see that

$$y_1 = \frac{x_1 + x_2}{2}$$

and

$$y_2 = \frac{x_1 - x_2}{2}.$$

These could be interpreted as the frequency components of $\mathbf{x}$, with y_1 representing the DC component and y_2 representing the high-frequency component. With this interpretation, the matrix A maps points in a spatial domain to points in a frequency domain. Most discrete transforms, such as the discrete Fourier transform examined in Chap. 2 and the discrete cosine transform used in JPEG and MPEG compression, can be represented as matrix transformations.

Consider the grades example in Fig. 4.5. We are often interested in finding students whose grades are erratic. In the frequency domain, erratic grades would result in larger magnitudes for high-frequency coefficients. Let's take the data matrix from the situation in Fig. 4.5 and map it into the frequency domain using a Discrete Walsh-Hadamard Transform. The four-dimensional Discrete Walsh Hadamard matrix

$$A = \frac{1}{4}\begin{bmatrix} 1 & 1 & 1 & 1 \\ 1 & 1 & -1 & -1 \\ 1 & -1 & -1 & 1 \\ 1 & -1 & 1 & -1 \end{bmatrix}$$

provides a mapping that generates a frequency representation of a four-dimensional vector of values $\mathbf{x}$. For Alice, the frequency coefficients are

$$\frac{1}{4}\begin{bmatrix} 1 & 1 & 1 & 1 \\ 1 & 1 & -1 & -1 \\ 1 & -1 & -1 & 1 \\ 1 & -1 & 1 & -1 \end{bmatrix}\begin{bmatrix} 9 \\ 4 \\ 10 \\ 8 \end{bmatrix} = \begin{bmatrix} 7.75 \\ -1.25 \\ 0.75 \\ 1.75 \end{bmatrix}$$

The first coefficient, 7.75, represents the DC coefficient of Alice's grades, which is the average value. The magnitude of the high-frequency coefficient is 1.75. To understand what this means, let's compare the coefficients corresponding to Alice's grades with the coefficients corresponding to Borges' grades

$$\frac{1}{4}\begin{bmatrix} 1 & 1 & 1 & 1 \\ 1 & 1 & -1 & -1 \\ 1 & -1 & -1 & 1 \\ 1 & -1 & 1 & -1 \end{bmatrix}\begin{bmatrix} 10 \\ 1 \\ 10 \\ 2 \end{bmatrix} = \begin{bmatrix} 5.75 \\ -0.25 \\ 0.25 \\ 4.25 \end{bmatrix}$$

The high-frequency coefficient for Borges is more than twice that of Alice. Looking at their test grades, we can see why. Borges has test grades that fluctuate wildly. Compare this to David's grades.

$$\frac{1}{4}\begin{bmatrix} 1 & 1 & 1 & 1 \\ 1 & 1 & -1 & -1 \\ 1 & -1 & -1 & 1 \\ 1 & -1 & 1 & -1 \end{bmatrix}\begin{bmatrix} 3 \\ 4 \\ 8 \\ 8 \end{bmatrix} = \begin{bmatrix} 5.75 \\ -2.25 \\ -0.25 \\ -0.25 \end{bmatrix}$$

David's average grade is the same as Borges's. So if we were only to look at average grades, we wouldn't find any difference between Borges and David. The difference becomes very clear when we look at the high-frequency coefficient. The magnitude of the high-frequency coefficient for Borges is seventeen times that of the magnitude of the high-frequency coefficient for David. There is a difference between these two students, requiring very different interventions.

By organizing the grades in a matrix and then using a mapping, we obtained a handy feature. We might have been able to construct this feature without using the matrix framework, but having the framework available made this particular tool immediately accessible.

If the columns of the matrix are independent, another way to look at the effect of matrix multiplication is to view the result as an expansion in terms of the columns of the matrix. A four-dimensional vector $\mathbf{x}$ with components x_1, x_2, x_3, and x_4 can be written as

$$\mathbf{x} = \begin{bmatrix} x_1 \\ x_2 \\ x_3 \\ x_4 \end{bmatrix} = x_1\begin{bmatrix} 1 \\ 0 \\ 0 \\ 0 \end{bmatrix} + x_2\begin{bmatrix} 0 \\ 1 \\ 0 \\ 0 \end{bmatrix} + x_3\begin{bmatrix} 0 \\ 0 \\ 1 \\ 0 \end{bmatrix} + x_4\begin{bmatrix} 0 \\ 0 \\ 0 \\ 1 \end{bmatrix}$$

The components of $\mathbf{x}$ are the coefficients of the unit vectors. Now, consider the transformation $\mathbf{y} = A\mathbf{x}$.

$$\mathbf{y} = \begin{bmatrix} a_{11} & a_{12} & a_{13} & a_{14} \\ a_{21} & a_{22} & a_{23} & a_{24} \\ a_{31} & a_{32} & a_{33} & a_{34} \\ a_{41} & a_{42} & a_{43} & a_{44} \end{bmatrix}\begin{bmatrix} x_1 \\ x_2 \\ x_3 \\ x_4 \end{bmatrix} = \begin{bmatrix} a_{11}x_1 + a_{12}x_2 + a_{13}x_3 + a_{14}x_4 \\ a_{21}x_1 + a_{22}x_2 + a_{23}x_3 + a_{24}x_4 \\ a_{31}x_1 + a_{32}x_2 + a_{33}x_3 + a_{34}x_4 \\ a_{41}x_1 + a_{42}x_2 + a_{43}x_3 + a_{44}x_4 \end{bmatrix}$$

$$= x_1\begin{bmatrix} a_{11} \\ a_{21} \\ a_{31} \\ a_{41} \end{bmatrix} + x_2\begin{bmatrix} a_{12} \\ a_{22} \\ a_{32} \\ a_{42} \end{bmatrix} + x_3\begin{bmatrix} a_{13} \\ a_{23} \\ a_{33} \\ a_{43} \end{bmatrix} + x_4\begin{bmatrix} a_{14} \\ a_{24} \\ a_{34} \\ a_{44} \end{bmatrix}$$

In other words, the vector $\mathbf{y}$ is simply the vector $\mathbf{x}$ expressed using a different basis set. The vectors $\mathbf{x}$ and $\mathbf{y}$ are two different representations of the same set of data values. By picking different sets of basis vectors, we can obtain different representations of the data values. Different representations can lead to features that are better suited to the task of separating the dataset into classes.

4.2.3 A Matrix is a Rotational Operator

A different interpretation of matrices is that matrices are rotational operators. Consider the matrix

$$A = \begin{bmatrix} 1 & 2 \\ 0 & 1 \end{bmatrix}$$

If this operates on the vector

$$\mathbf{x} = \begin{bmatrix} 1 \\ 2 \end{bmatrix}$$

we get

$$\mathbf{y} = \begin{bmatrix} 1 & 2 \\ 0 & 1 \end{bmatrix} \begin{bmatrix} 1 \\ 2 \end{bmatrix} = \begin{bmatrix} 5 \\ 2 \end{bmatrix}$$

Graphically, we can show this as in Fig. 4.6. We can see that the black vector corresponding to **x** has been rotated and stretched to generate the blue vector corresponding to **y**. If we set

$$\mathbf{x} = \begin{bmatrix} 2 \\ 2 \end{bmatrix}$$

we get the result shown in Fig. 4.7

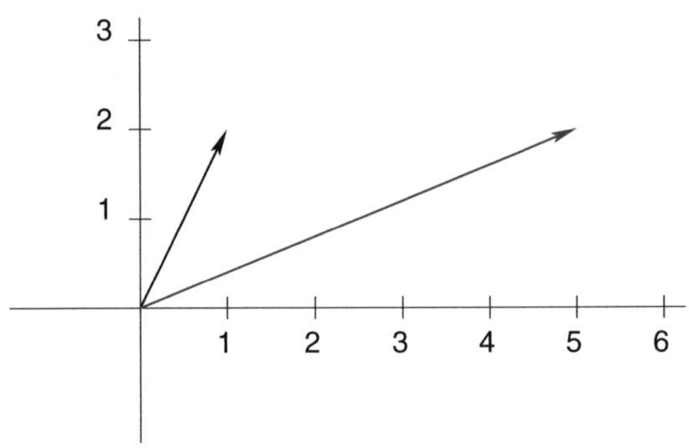

Fig. 4.6 Rotation of the vector $\mathbf{x} = [1\ \ 2]^T$ using a matrix operation to get the vector $\mathbf{y} = [5\ \ 2]^T$

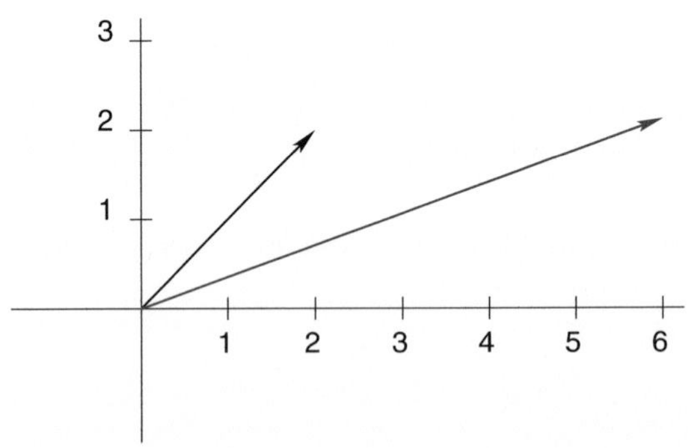

Fig. 4.7 Rotation of the vector $\mathbf{x} = [2\ \ 2]^T$ using a matrix operation to get the vector $\mathbf{y} = [6\ \ 2]^T$

What if

$$\mathbf{x} = \begin{bmatrix} 2 \\ 0 \end{bmatrix}$$

then

$$\mathbf{y} = \begin{bmatrix} 1 & 2 \\ 0 & 1 \end{bmatrix} \begin{bmatrix} 2 \\ 0 \end{bmatrix} = \begin{bmatrix} 2 \\ 0 \end{bmatrix}$$

This particular vector is rotationally invariant! Such a vector is called the *eigenvector* of the matrix. In general if we have a vector $\mathbf{v}$ such that

$$\mathbf{A}\mathbf{v} = \lambda \mathbf{v}$$

where λ is a scalar, then $\mathbf{v}$ is called the eigenvector of the matrix $\mathbf{A}$ and λ is called the eigenvalue. In terms of our understanding of the matrix as a rotational operator, an eigenvector of a matrix is a vector whose direction remains unchanged by the mapping operation performed by the matrix. The only thing that can change is the magnitude, which is reflected in the eigenvalue. In this particular case, the eigenvector is $[2\ 0]^T$, and the eigenvalue is 1.

We will use this view of the matrix as a rotational operator in the next section.

4.3 Matrix Factorization

Now that we have some idea of different ways of interpreting matrices, let's look at how matrices can be used to generate features. One way, of course, is through transformations. Using a frequency transformation, we can obtain frequency coefficients that serve as features. We showed one example in the previous section. Another popular application is determining whether a specific region of an image contains edges. An edge in an image is a region where the pixel values change significantly. The amount of change in the data can be understood in terms of frequency; significant changes correspond to higher frequencies. Therefore, we could find the frequency coefficients corresponding to the pixel values of the image in that region. We could then use the magnitude of the high-frequency coefficient to decide whether the region contained edges.

Matrix factorization is another method for exploring the hidden characteristics of the data. To do this, we first have to organize our data into a matrix. We have already seen one example where we organized student grades into a matrix. For a different example, consider a scenario where we have 8,000 samples of speech from twelve different speakers. We could construct a 12×8000 matrix, where each row contains the speech samples for a different speaker. Alternatively, we could find a high-order model for each speaker, which would constitute the rows of the matrix. In each case, the rows correspond to different objects, in this case, speakers, and the columns correspond to attributes. These attributes would be samples of speech if each row contained the speech samples, or they would be the model coefficients if each row contained the autoregressive model for the speech. Once we have set up the $n \times m$ matrix **A**, where n is the number of objects and m is the number of attributes the factorization consists of writing **A** as a product of three matrices

$$\mathbf{A}_{n\times m} = \mathbf{C}_{n\times r}\mathbf{W}_{r\times r}\mathbf{F}_{r\times m}$$

There are several ways to factor a matrix, and several ways to interpret the factorization. Different matrix factorizations make different assumptions about the underlying structure present in the data. For example, we can view factorization as a means of revealing hidden factors. While the rows of **A** consist of the attributes of this data, these might be mixtures of latent attributes of particular significance. The latent attributes are columns in **F** and the mixture components are rows in **C**. We can also interpret factorization as a method for revealing hidden clusters. This might seem needless. When we examine a plot of data points, we can usually identify the clusters. However, the geometry of higher-dimensional space is not intuitive, and it might not be that easy to see how the data is clustered. In an application where we are looking for clusters, each row of **F** represents an axis, and the rows of **C** correspond to the coordinates of the data points. Elements of (a diagonal) **W** cause different degrees of stretching of the axes. There are several other possible ways of looking at **A** and corresponding methods of factorization. One widespread factorization technique is *Singular Value Decomposition*.

4.3.1 Singular Value Decomposition

When we select features, such as height, weight, strength, and age in our football example, we can organize these features into vectors. It seems that having more features would be more conducive to obtaining a good classifier. And in many ways, this is true. However, as we increase the size of the vector, we also increase the dimension of the space in which we are operating, which can lead to several problems, such as numerical instability, increased computational cost, and overfitting. Luckily, there is a way to have our cake and eat it too—increase the number of features without increasing the dimensionality of the space we have to operate in. The features that we select are often redundant; if we have two features, adding two more does not necessarily double the amount of information we previously possessed. This is not to say that the additional features do not contribute to our total stock of information—just that some of the information we obtain could have been inferred from the previously available data. For example, consider the features of height and weight. If we know someone's height, we have an idea of what their weight is. Therefore, knowing their weight does not double the amount of information we had when we only knew their height. We cannot simply discard one of these features; however, we can explore combining them to create a new synthesized feature that retains most of the information from both. If this combination is a linear combination, then this means we need to come up with a matrix that will map this two-dimensional feature vector into a one-dimensional synthetic feature, which can be used for our classification task.

Consider the following toy problem. We are given the heights and weights of football players shown below.

Height (cm)	Weight (kg)
180	95
182	90
184	90
182	100
188	105
190	100
190	110
194	110
196	115
196	135
196	140
200	142

The task is to assign the players to different positions. There is, of course, too little information for the task at hand, but we could use some subjective rules: quarterbacks will probably be lighter, running backs will be shorter and lighter, while tackles and guards will be bigger, both in terms of height and weight. Therefore, the classification task can be accomplished by partitioning the two-dimensional height-weight space into regions, with each region corresponding to a particular position. Can we simplify the classification task further?

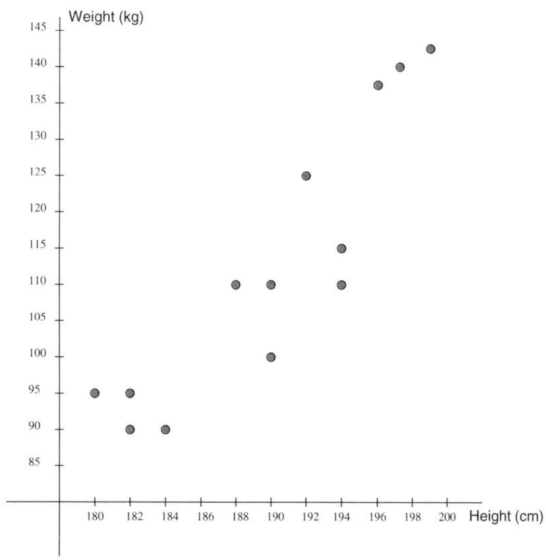

Fig. 4.8 Heights and weights of football players

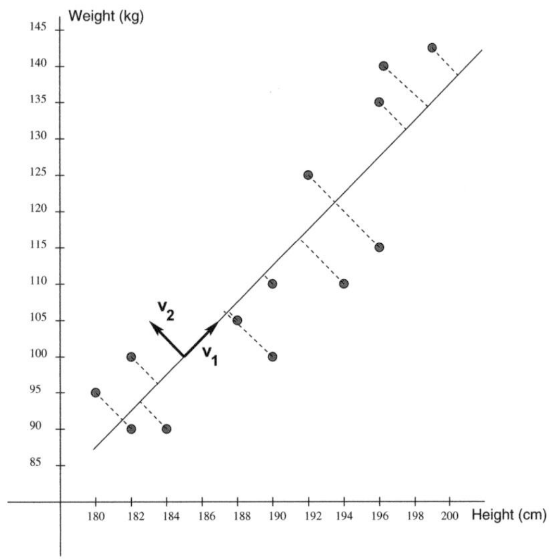

Fig. 4.9 Heights and weights of football players

If we plot this data as shown in Fig. 4.8, we can see that the height and weight are related. Larger values of height generally correspond to larger values of weight. In fact, we could draw a line as shown in Fig. 4.9 around which the feature vectors are clustered. What we can do now is to project the height-weight points onto this line and use the distance along this line for classification. That way, the problem of partitioning a two-dimensional space into regions gets reduced to partitioning a line into intervals. While this doesn't seem like a significant gain, the procedure we develop can also be applied to reduce the problem of partitioning a much higher-dimensional space, such as a ten-dimensional space, to the problem of partitioning a two-dimensional space. Let's continue. How do we project the points onto the line? We can do that by finding a unit vector along the line and then taking the dot product of the point with the unit vector. Let's pick two points on the line, say the point at coordinates (191, 115) and the point at coordinates (187, 105). The vector along the line can be obtained as

$$\begin{bmatrix} 191 \\ 115 \end{bmatrix} - \begin{bmatrix} 187 \\ 105 \end{bmatrix} = \begin{bmatrix} 4 \\ 10 \end{bmatrix}$$

The unit vector $\mathbf{v_1}$ can be obtained by normalizing this vector by its magnitude.

$$\mathbf{v}_1 = \frac{1}{\sqrt{116}} \begin{bmatrix} 4 \\ 10 \end{bmatrix}$$

We can obtain the projection of a feature vector by taking its dot product with $\mathbf{v}$. For example, the second feature vector in our list, with a height of 182cm and a weight of 90 kg, projects onto the point

$$\begin{bmatrix} 182 \\ 90 \end{bmatrix} \cdot \frac{1}{\sqrt{109}} \begin{bmatrix} 3 \\ 10 \end{bmatrix} = 138.5$$

while the feature vector corresponding to a height of 190 cm and a weight of 110 kg maps onto the point

$$\begin{bmatrix} 190 \\ 110 \end{bmatrix} \cdot \frac{1}{\sqrt{109}} \begin{bmatrix} 3 \\ 10 \end{bmatrix} = 160$$

Rather than obtain the projections one at a time, we can rotate the entire space and express the feature vectors in terms of the basis vectors $\mathbf{v_1}$ and $\mathbf{v_2}$ where $\mathbf{v_2}$ is orthogonal to $\mathbf{v_1}$. To do that, we need the basis vector $\mathbf{v_2}$. This is easy to do in this case. Suppose, $\mathbf{v_2} = [a\ b]^T$ then because $\mathbf{v_1}$ and $\mathbf{v_2}$ are orthogonal we get

$$\frac{1}{\sqrt{109}} \begin{bmatrix} 3 \\ 10 \end{bmatrix} \cdot \begin{bmatrix} a \\ b \end{bmatrix} = \frac{1}{\sqrt{109}}(3a + 10b) = 0$$

We only have one equation and two unknowns. The reason is that two directions are orthogonal to $\mathbf{v_1}$, one pointing up and one pointing down. Additionally, we have not utilized the fact that $\mathbf{v_2}$ has a magnitude of one, i.e., $a^2 + b^2 = 1$. If we pick the direction of $\mathbf{v_2}$ shown in the figure an easy choice is $b = 3$, which means $a = -10$ and

$$\mathbf{v_2} = \frac{1}{\sqrt{109}} \begin{bmatrix} -10 \\ 3 \end{bmatrix}$$

and the rotation matrix R becomes

$$R = \frac{1}{\sqrt{109}} \begin{bmatrix} 3 & -10 \\ 10 & 3 \end{bmatrix}$$

Rotating all the feature vectors, we obtain the situation shown in Fig. 4.10.

This was a simple two-dimensional case. How can we generalize this to higher dimensions? We drew the line on which we obtained projections by simply eyeballing it. Was it the best line? What does best mean? We need to develop a systematic method for obtaining the projections. This is what we will now do. In the process, we will do a hand-waving derivation of the Singular Value Decomposition. (The description (though not the hand waving) is based on an unpublished book by Hopcroft and Kannan. At this writing, you can find the book at https://www.cs.cmu.edu/~venkatg/teaching/CStheory-infoage/hopcroft-kannan-feb2012.pdf.)

If you are not interested in our hand-waving derivation, here is the punchline: organize the feature vectors into a matrix $\mathbf{A}$. The basis vectors we are after are the eigenvectors of $\mathbf{A}^T\mathbf{A}$.

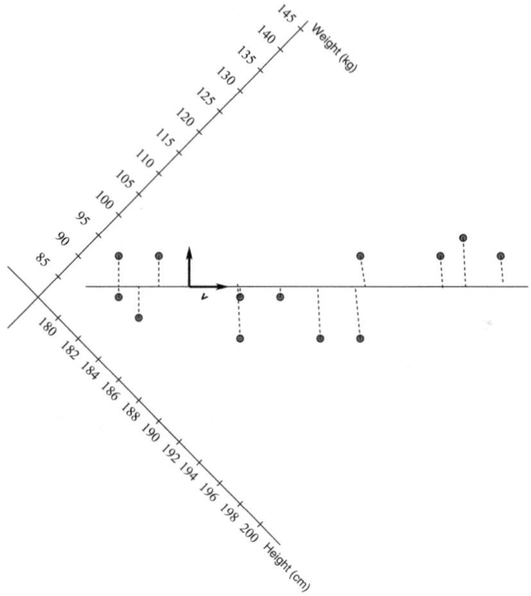

Fig. 4.10 Heights and weights of football players

An Approximation Approach

Suppose we have a set of points $\{\mathbf{x}_i\} \in \mathcal{R}^k$. Each of these points can be viewed as a vector in $\mathcal{R}^k$. A point in k-dimensional space can be represented in multiple ways by using k independent vectors $\{\mathbf{v}_i\}$. It is most convenient if these vectors are orthonormal because then we can find the component in the direction of $\mathbf{v}_i$ by simply taking the dot product of the data point with $\mathbf{v_i}$. What we would like to do is, given a set of data points $\{\mathbf{x}_i\}$, to find a basis set $\{\mathbf{v}_i\}$ such that the representation with respect to this basis set is most informative. Consider the set of data points in Fig. 4.11. We can see that the values of the data vary most along a line. It is reasonable to assume that most of the information contained in the data will be along this line. So, we will pick our first basis vector to be in this direction. We can achieve this by finding the best-fitting line, which is defined as the line that minimizes the average distance of all our data points to the line. We will define $\mathbf{v}_1$ to be the unit vector corresponding to this line.

The best fitting line in the mean squared sense would be the line to which the distances of the points are minimized (Fig. 4.12).

Taking a look at one of these points (Fig. 4.13) and looking at its relationship to the line. To better see the relationship we have rotated the portion in the circle. We can see that the projection of the data point $\mathbf{x_i}$ onto the line b forms a right-angled triangle. We know the distances b, p, and $|\mathbf{x_i}|$ are related through Pythagoras' theorem:

$$b^2 + p^2 = |\mathbf{x_i}|^2$$

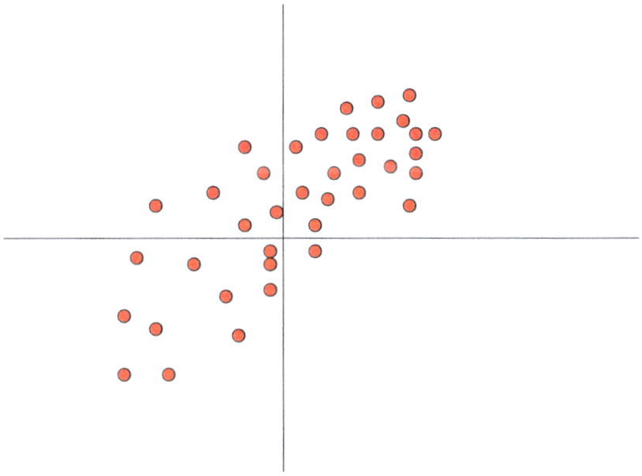

Fig. 4.11 Data in two dimensions

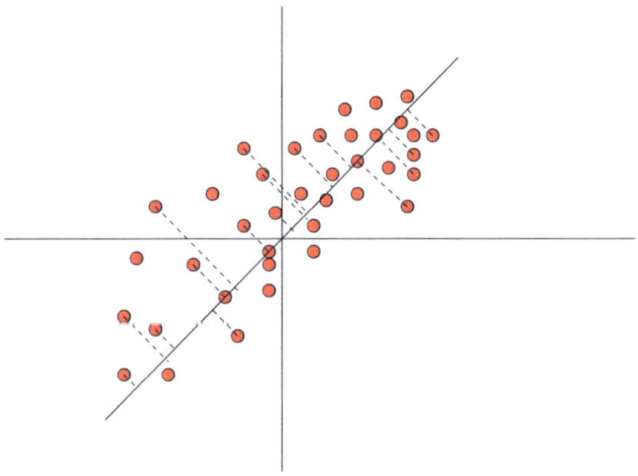

Fig. 4.12 The best fitting line to the 2D data

or

$$p^2 = |\mathbf{x_i}|^2 - b^2$$

What we want to do is find a line such that the distance of the point to the line, p, is minimized. As $\mathbf{x_i}$ is our feature vector, its value is fixed. Therefore, if we want to minimize p^2, we have to maximize b^2 while keeping b perpendicular to p.

If we now include a unit vector in the direction of the line the fact that b is a projection of $\mathbf{x_i}$ onto the line can be used to write b as:

$$b = \mathbf{x_i} \cdot \mathbf{v}$$

Recall that $\mathbf{x_i}$ is a row of the $\mathbf{A}$ matrix. What we would like to do is find a vector $\mathbf{v}$ which minimizes the sum of the distances of all the data points we have (Fig. 4.14), or in other words, for all the rows of $\mathbf{A}$. As the ith row of $\mathbf{A}$ is $\mathbf{x}_i$ then this would be

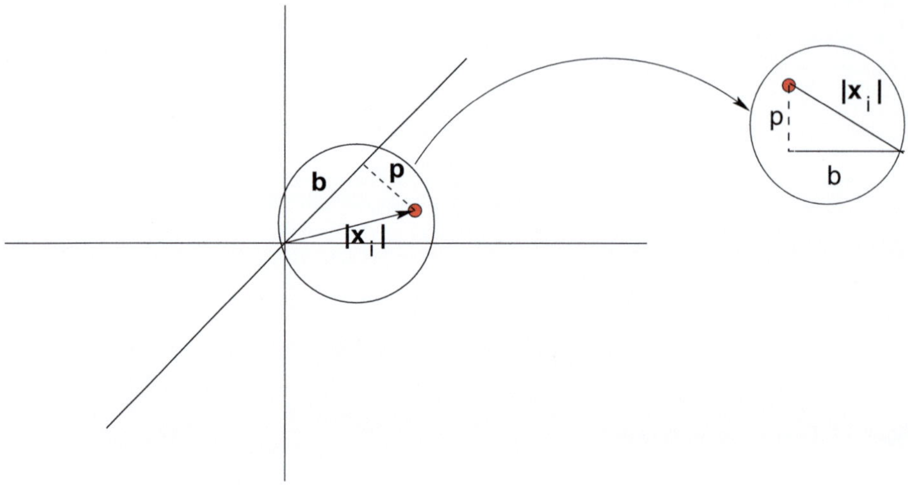

Fig. 4.13 One point from the data vector

Fig. 4.14 2D data

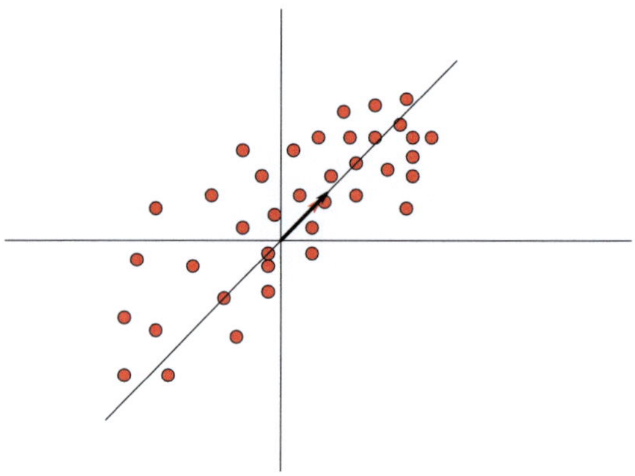

$$\sum_i (\mathbf{x}_i \cdot \mathbf{v})^2$$

Or, expanding this,

$$\sum_i (\mathbf{x}_i \cdot \mathbf{v})^2 = (\mathbf{x}_1 \cdot \mathbf{v})^2 + (\mathbf{x}_2 \cdot \mathbf{v})^2 + \cdots + (\mathbf{x}_n \cdot \mathbf{v})^2$$

Because

$$A = \begin{bmatrix} \mathbf{x}_1 \\ \mathbf{x}_2 \\ \vdots \\ \mathbf{x}_n \end{bmatrix}$$

so

$$A \cdot \mathbf{v} = \begin{bmatrix} \mathbf{x}_1 \cdot \mathbf{v} \\ \mathbf{x}_2 \cdot \mathbf{v} \\ \vdots \\ \mathbf{x}_n \cdot \mathbf{v} \end{bmatrix}$$

and $|\mathbf{A} \cdot \mathbf{v}|^2$ is given by the

$$|A \cdot \mathbf{v}|^2 = (\mathbf{x}_1 \cdot \mathbf{v})^2 + (\mathbf{x}_2 \cdot \mathbf{v})^2 + \cdots + (\mathbf{x}_n \cdot \mathbf{v})^2$$

or

$$\sum_i (\mathbf{x}_i \cdot \mathbf{v})^2 = |\mathbf{A}\mathbf{v}|^2$$

Therefore, the best-fit line will be one that maximizes $|\mathbf{A}\mathbf{v}|^2$ or $|\mathbf{A}\mathbf{v}|$.

$$\mathbf{v}_1 = \arg \max_{|\mathbf{v}|=1} |\mathbf{A}\mathbf{v}|$$

The value $|\mathbf{A}\mathbf{v}_1|$ is called the *first singular value* of $\mathbf{A}$ and is often denoted by $\sigma_1(\mathbf{A})$. The singular value is the component of $\mathbf{A}$ along $\mathbf{v}_1$.

To find $\mathbf{v}_2$ such that $\mathbf{v}_1$ and $\mathbf{v}_2$ span the two-dimensional subspace that best fits the data, we use

$$\mathbf{v}_2 = \arg \max_{\mathbf{v}_2 \perp \mathbf{v}_1, |\mathbf{v}|-1} |\mathbf{A}\mathbf{v}|$$

$$\sigma_2(\mathbf{A}) = |\mathbf{A}v_2|$$

is the second singular value of $\mathbf{A}$.

Rather than sequentially finding $\mathbf{v}_1$ and $\mathbf{v}_2$, could we have done better if we had looked jointly for $\mathbf{v}_1$ and $\mathbf{v}_2$? For the metric we are using to determine our best fit, namely the sum of squared distances, the sequential or greedy approach works just as well as a more global approach. Continuing in this manner we can obtain the right singular vectors of $\mathbf{A}$, $\{\mathbf{v}_1, \mathbf{v}_2 \ldots, \mathbf{v}_r\}$, where r is the rank of $\mathbf{A}$, or the dimension of the space in which the data lives. We have now systematically formulated our problem, resulting in a set of basis vectors where the first basis vector points in the most informative direction, the second in the second most informative direction, and so on. This means that if we had these basis vectors, we could find the most informative reduced feature vector by projecting our feature vector onto the most informative basis vectors. However, we don't yet know how to find these basis vectors. To achieve that, we have a bit more work to do.

Define

$$\mathbf{u}_i = \frac{1}{\sigma_i(\mathbf{A})}\mathbf{A}\mathbf{v}_i$$

These are called the left singular vectors of $\mathbf{A}$, and we can show that $\{\mathbf{u}_i\}$ are orthogonal.

Let's define $\mathbf{U}$ as the matrix with rows $\mathbf{u}_i$ and $\mathbf{V}$ as the matrix with rows $\mathbf{v}_j$. Then our claim is that

$$\mathbf{A} = \sum_{i=1}^{r} \sigma_i \mathbf{u}_i \mathbf{v}_i^T = \mathbf{U}\Sigma\mathbf{V}^T$$

We will prove our claim by making use of the fact that if for two matrices **A** and **B**

$$\mathbf{A}\mathbf{v} = \mathbf{B}\mathbf{v}$$

for all arbitrary **v**, then $\mathbf{A} = \mathbf{B}$.

Any arbitrary vector **v** can be represented by a linear combination of $\{\mathbf{v}_1, \mathbf{v}_2 \ldots, \mathbf{v}_r\}$, which span the space in which our data lives, and a vector **w** which is orthogonal to the space spanned by $\{\mathbf{v}_1, \mathbf{v}_2 \ldots, \mathbf{v}_r\}$. In other words, we can express an arbitrary vector **v** as

$$\mathbf{v} = \sum_{j=1}^{r} \alpha_j \mathbf{v}_j + \alpha_w \mathbf{w}$$

Then,

$$\mathbf{A}\mathbf{v} = \mathbf{A}(\sum_{j=1}^{r} \alpha_j \mathbf{v}_j + \alpha_w \mathbf{w}) = \sum_{j=1}^{r} \alpha_j \mathbf{A}\mathbf{v}_j + \alpha_w \mathbf{A}\mathbf{w}$$

The last term **Aw** is zero as **w** is orthogonal to the space in which **A** lives. Therefore,

$$\mathbf{A}\mathbf{v} = \sum_{j=1}^{r} \alpha_j \mathbf{A}\mathbf{v}_j$$

From the definition of $\mathbf{u}_i$ we know that

$$\mathbf{A}\mathbf{v}_i = \sigma_i \mathbf{u}_i$$

Also because the vectors $\{\mathbf{v}_i\}$ are orthonormal

$$\mathbf{v}_i \cdot \mathbf{v}_j = \mathbf{v}_i^T \mathbf{v}_j = \begin{cases} 1 & i = j \\ 0 & i \neq j \end{cases}$$

Using these two facts we can write $\mathbf{A}\mathbf{v}_j$ as

$$\mathbf{A}\mathbf{v}_j = \sum_{i=1}^{r} \sigma_i \mathbf{u}_i \mathbf{v}_i^T \mathbf{v}_j$$

The reason we can do this is because all terms in this summation for which $i \neq j$ will be zero because $\mathbf{v}_i^T \mathbf{v}_j$ is zero when $i \neq j$. The only non-zero term is the term for which $i = j$, and for this term $\mathbf{v}_i^T \mathbf{v}_j$ is one.

Replace $\mathbf{A}\mathbf{v}_j$ in the expression for **Av** we get

$$\mathbf{A}\mathbf{v} = \sum_{j=1}^{r} \alpha_j \sum_{i=1}^{r} \sigma_i \mathbf{u}_i \mathbf{v}_i^T \mathbf{v}_j = \sum_{i=1}^{r} \sigma_i \mathbf{u}_i \mathbf{v}_i^T \sum_{j=1}^{r} \alpha_j \mathbf{v}_j$$

We can add $\mathbf{v}_i^T \mathbf{w}$ for free because it will be zero for all i, so

$$\mathbf{A}\mathbf{v} = \sum_{i=1}^{r} \sigma_i \mathbf{u}_i \mathbf{v}_i^T \left(\sum_{j=1}^{r} \alpha_j \mathbf{v}_j + \alpha_w w \right) = \sum_{i=1}^{r} \sigma_i \mathbf{u}_i \mathbf{v}_i^T \mathbf{v}$$

As $\mathbf{v}$ was arbitrary, this is true for all $\mathbf{v}$ and

$$\mathbf{A} = \sum_{i=1}^{r} \sigma_i \mathbf{u}_i \mathbf{v}_i^T = U \Sigma V^T$$

So, how do we find these vectors? Let $\mathbf{B} = \mathbf{A}\mathbf{A}^T$

$$\begin{aligned}
\mathbf{B} &= \left(\sum_{i=1}^{r} \sigma_i \mathbf{u}_i \mathbf{v}_i^T \right) \left(\sum_{j=1}^{r} \sigma_j \mathbf{u}_j \mathbf{v}_j^T \right)^T \\
&= \sum_{i=1}^{r} \sum_{j=1}^{r} \sigma_i \mathbf{u}_i \mathbf{v}_i^T \mathbf{v}_j \mathbf{u}_j^T \sigma_j \\
&= \sum_{i=1}^{r} \sigma_i^2 \mathbf{u}_i \mathbf{u}_i^T
\end{aligned}$$

Now

$$\mathbf{B}\mathbf{u}_j = \sum_{i=1}^{r} \sigma_i^2 \mathbf{u}_i \mathbf{u}_i^T \mathbf{u}_j = \sigma_j^2 \mathbf{u}_j$$

which means that $\mathbf{u}_j$ is an eigenvector of $\mathbf{B}$ or $\mathbf{A}\mathbf{A}^T$ and σ_j^2 is the corresponding eigenvalue.

Now define $\mathbf{C} = \mathbf{A}^T\mathbf{A}$

$$\begin{aligned}
\mathbf{C} &= \left(\sum_{i=1}^{r} \sigma_i u_i v_i^T \right)^T \left(\sum_{j=1}^{r} \sigma_j u_j v_j^T \right) \\
&= \sum_{i=1}^{r} \sum_{j=1}^{r} \sigma_i v_i u_i^T u_j v_j^T \sigma_j \\
&= \sum_{i=1}^{r} \sigma_i^2 v_i v_i^T
\end{aligned}$$

Then

$$\mathbf{C}\mathbf{v}_j = \sum_{i=1}^{r} \sigma_i^2 \mathbf{v}_i \mathbf{v}_i^T \mathbf{v}_j = \sigma_j^2 \mathbf{v}_j$$

or $\mathbf{v}_j$ is the eigenvector of $\mathbf{A}^T\mathbf{A}$ and σ_j^2 is the corresponding eigenvalue.

So to find the vectors $\mathbf{v}_j$ all we need to do is to stick our feature vectors into a matrix $\mathbf{A}$, find $\mathbf{A}^T\mathbf{A}$ and then find the eigenvalues of $\mathbf{A}^T\mathbf{A}$.

The Singular Value Decomposition gives us a set of vectors in which the first vector corresponds to the most informative direction of the data, the second vector is the next most informative, and so on. If we now want to remove the least informative values of the data and keep only the most informative values all we need to do is to delete the vectors corresponding to the least informative values in $\mathbf{V}$ and then multiply the data matrix by this reduced set of basis vectors and we will obtain a reduced and hopefully more informative representation of the data. This allows us to reduce the dimensionality of the data to only the significant values. In a sense, we are removing components of the data that had the effect of diluting the main correlations.

MATLAB has a function *svd*, which provides the singular value decomposition of a matrix. Similar functions are available in open-source alternatives, such as Python, Octave, and Julia, among others. Let's use the MATLAB function to look at our grade data. We first set up the data matrix

```
>> A = [9 4 10 8;10 1 10 2; 10 5 10 10; 3 4 7 8; 2 5 4 10]

A =

     9     4    10     8
    10     1    10     2
    10     5    10    10
     3     4     7     8
     2     5     4    10
```

The command for singular value decomposition is

```
>> [U S V] = svd(A)
```

where U and V are the eigenvalue matrices described above and S is the matrix of eigenvalues. When we use this command with our matrix, we get the following results:

```
U =

   -0.5172   -0.1066   -0.0307    0.6553    0.5393
   -0.4036   -0.6882    0.2232   -0.5406    0.1465
   -0.5772    0.0131   -0.5386    0.0259   -0.6131
   -0.3591    0.3195    0.7976    0.0953   -0.3517
```

```
   -0.3278    0.6425   -0.1518   -0.5183    0.4337


S =

   31.1707         0         0         0
         0    9.9077         0         0
         0         0    2.0551         0
         0         0         0    0.0000
         0         0         0         0


V =

   -0.5196   -0.5517   -0.6524         0
   -0.2706    0.3474   -0.0783   -0.8944
   -0.6033   -0.3038    0.7374    0.0000
   -0.5411    0.6947   -0.1565    0.4472
```

Notice that while the grade vectors are four-dimensional, there are only three non-zero eigenvalues. The reason becomes evident when we examine the relationship between the grades for Test 2 and Test 4. For each student, the grade for Test 4 is twice the grade for Test 2. There are only three independent values in the grade vectors. Hence, the dimensionality of the data can be reduced from four to three without any loss of information.

The matrix V contains the basis vectors for the reduced-dimensionality space. To obtain the representations of the grade vectors in this space, we simply multiply the data matrix A with the basis vectors.

```
>> A*V

ans =

  -16.1208   -1.0557   -0.0630    0.0000
  -12.5819   -6.8180    0.4586    0.0000
  -17.9933    0.1294   -1.1068    0.0000
  -11.1933    3.1659    1.6392    0.0000
  -10.2166    6.3658   -0.3120         0
```

We now have three-dimensional representations of the four-dimensional grade vectors. Even in this reduced dimensionality, the coefficients in the third dimension are relatively small. Zeroing out the third dimension, we can plot the two-dimensional representation as

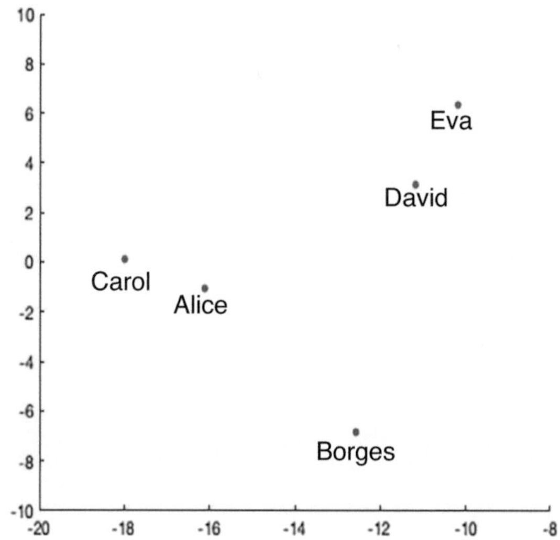

Fig. 4.15 Scatterplot of the reduced dimensional representation of the grade vectors

shown in Fig. 4.15. From the scatterplot, we can see that Carol and Alice form one group, Eva and David form another, and Borges is an outlier.

This was a toy example, and we could probably have obtained the same insights by examining the raw data. However, we can see the power of this approach when we consider that many datasets in current biomedical applications consist of thousands of high-dimensional data vectors. Singular value decomposition offers a straightforward and efficient method for gaining insights into the data.

4.3.2 Principal Component Analysis

Principal component analysis is essentially the last part of our description of the SVD. The idea is to come up with a representation of the data in terms of its "principal" axes, which are the axes with the largest variance in the data. Principal Component Analysis operates on data that has been mean-removed. By mean-removed data, we mean we subtract the mean value of each column. In the case of our toy example, the mean-removed data matrix would be

$$\bar{\mathbf{A}} = \begin{bmatrix} 2.2 & 0.2 & 1.8 & 0.4 \\ 3.2 & -2.8 & 1.8 & -5.6 \\ 3.2 & 1.2 & 1.8 & 2.4 \\ -3.8 & 0.2 & -1.2 & 0.4 \\ -4.8 & 1.2 & -4.2 & 2.4 \end{bmatrix}$$

Denoting the mean removed data matrix as $\bar{\mathbf{A}}$, we first obtain the normalized eigenvectors of $\bar{\mathbf{A}}\bar{\mathbf{A}}^T$ and organize them in the matrix $\mathbf{V}$. We then obtain the representation of the mean-

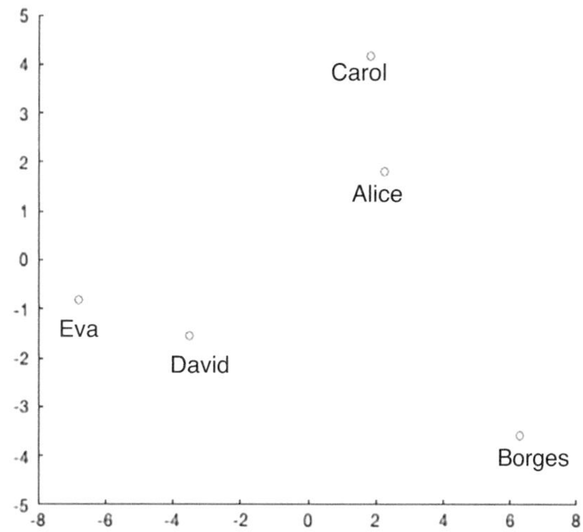

Fig. 4.16 Scatterplot of the principal components of the grade vectors

removed data with respect to these vectors by multiplying $\bar{\mathbf{A}}$ by $\mathbf{V}$

$$\mathbf{C} = \bar{\mathbf{A}}\mathbf{V}$$

We can then zero out the least significant components of and keep only the principal components to be used in the analysis to follow.

Let's find and plot the principal components for our toy example. In MATLAB, we can either use the *svd* function with $\bar{\mathbf{A}}$ to find the basis set for the principal components, or we can use the function *pca* directly on the original grade matrix. A plot of the two-dimensional principal component vectors is shown in Fig. 4.16. Comparing this to Fig. 4.15, we can see that Principal Component Analysis provided a different basis set, but the relative positioning of the students remains the same. Carol and Alice group together, Eva and David group together, and Borges is an outlier.

4.4 Summary

A significant challenge in biomedical signal processing is classification, which involves assigning a signal to one of several possible classes. To determine whether a signal belongs to a particular class, we utilize the properties or features of the signal. Often, these features are application-dependent. However, there are also features that are application agnostic. There are two broad classes of popular application-agnostic features: statistical features and matrix-derived features. If the signals we are looking at are functions of time, we can treat them as realizations of a random process and extract the statistics, or statistical features, from these signals. If, on the other hand, the signals we are looking at are collections of discrete values, such as medical test results or physical parameters, then we can organize this data into matrices and extract matrix-dependent features from these matrices. In this chapter, we have examined several popular statistical features and matrix-derived features. These are not the only data-agnostic features we could have obtained, but they give us a sense of the kinds of features we can use when faced with a classification task.

Feature Classification

5

Now that we have several methods for generating features from our data, we need a way to extract useful information from a vector of features. One possible approach is to take the set of feature vectors and examine if subsets of the feature vectors fall into clusters in an informative manner. For example, we could obtain feature vectors corresponding to different organisms and see if they cluster according to their phylogenetic relationships. Such clustering would suggest that the features are related to evolutionary relationships. Another situation would be where we have a set of data that we know corresponds to distinct conditions or classes, and we want to ascertain which class a new data point belongs to. This process is called *classification*, which we turn our attention to next.

A classifier can be seen as a partitioning of the space in which the data lives. Consider our height-weight toy example. If we wanted to assign each individual to one of four classes, one possible classifier could be the partitioning of the two-dimensional space of height-weight points into four regions as shown in Fig. 5.1.

The problem of classifier design, then, is partitioning the data space. If our feature vectors had only two or three features, we could place them in a two- or three-dimensional space and visualize how to partition this space. However, when the number of features or the dimensionality of the feature vector becomes large, such an approach is no longer feasible, and we need algorithmic procedures.

Developing classifiers is something people have been doing for a long time, and nowadays, classifier development is a significant aspect of the field of machine learning. Covering the entire field is way beyond our scope. Instead, we will focus on two popular techniques that examine the problem from slightly different perspectives. The first algorithm we will look at is a popular clustering algorithm that incorporates class discovery. This algorithm is often referred to as the k-means algorithm but is also known as the Generalized Lloyd Algorithm.

K. Sayood, *Biomedical Signal Processing*, Synthesis Lectures on Biomedical Engineering,
https://doi.org/10.1007/978-3-032-08855-0_5

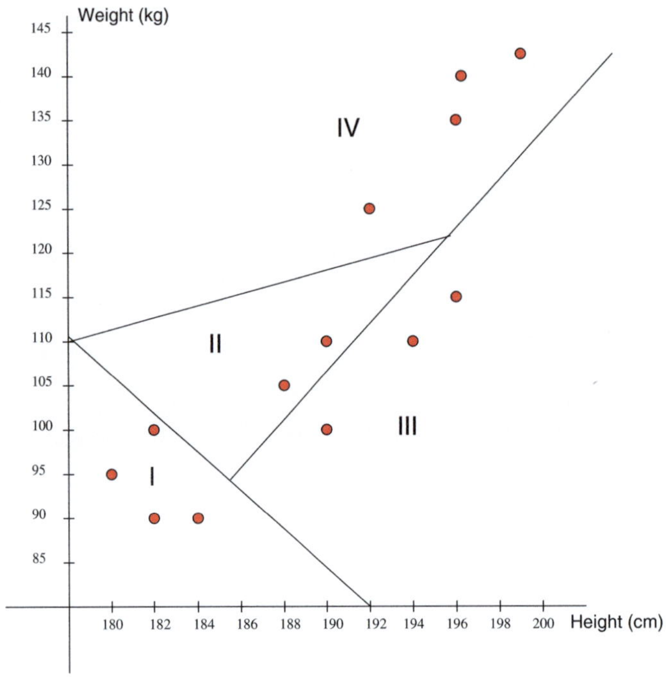

Fig. 5.1 A possible classifier which classifies the feature vector into one of four classes

The second is an algorithm in which, given a set of vectors belonging to different classes, we develop the optimum linear classifier called the *support vector machine*. This linear classifier can then be extended to nonlinear scenarios.

5.1 Generalized Lloyd Algorithm

Most adaptive algorithms–algorithms that try to adapt to a set of desiderata work on the general principle sketched out in Fig. 5.2. We have some data, a system/algorithm we are trying to design, and some desired responses from the system. We provide the data to the current iteration of the system, which generates an output. We then take the difference between the output of the current iteration of the system and our desired response. This difference, or error, is provided to the adaptation algorithm, which then modifies the parameters of the system with the goal (or hope) of driving the error to zero. We repeat this process until a stopping criterion is met. Many machine learning algorithms use this general approach. The idea is that the machine uses the difference between the desired response of the current iteration of the system and the actual response to learn what it needs to do and how to adjust the system's parameters in the next iteration.

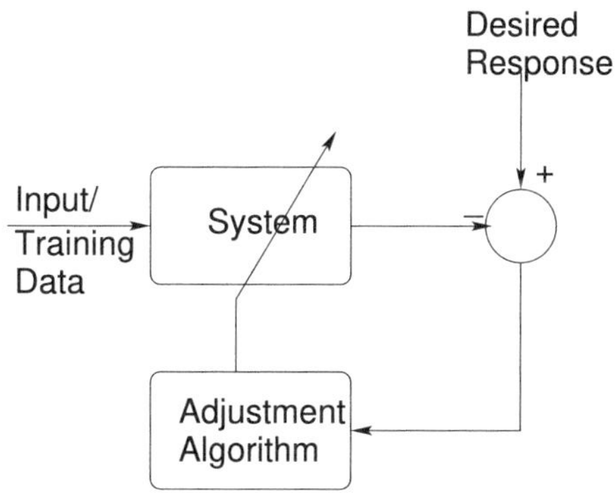

Fig. 5.2 A general framework for adaptive systems

In the case of clustering of feature vectors, we can define the system in one of two different ways. We can determine the boundaries of the classes such that when a feature vector falls within a particular set of boundaries, we say it belongs to the class specified by those boundaries. Another, more practical approach is to define representative vectors for each class. When we need to classify a particular feature vector, we compute the similarity or distance of the feature vector to the representative vectors and determine the appropriate class to be the one whose representative vector is most similar or closest to the feature vector. A popular method for picking the representative vector is to choose the vector that is the centroid of the corresponding region. In practice, we typically have training data sets that are either labeled as belonging to a particular class or that we assign to specific classes. If we have vectors belonging to a class, we can define the centroid of the region corresponding to that class as the average value of the vectors within that class. If we know the centroids of the clusters, the optimum partitions in the mean squared sense are the perpendicular bisectors of the line connecting the centroids. In other words, the region corresponding to a centroid is the set of points closer to that centroid than to any other centroid. Such a region is called a *Voronoi region*. The partitions themselves are mostly conceptual; the usual process for assigning a feature vector to a particular region is to compute the distance to the representative vectors and pick the region with the representative vector closest to the feature vector.

If we find the centroids of the data points in the four different regions in Fig. 5.1, we get the results shown in Fig. 5.3, where the centroids are shown in blue.

Given these centroids, however, our partitions do not define Voronoi regions. The next step is to adjust the partitions between regions so that we end up with Voronoi regions. When that happens, as shown in Fig. 5.4, we find that because the boundaries have shifted, the centroids we calculated no longer reflect the feature vectors in the various partitions.

Once the centroids are recomputed, we would then find new partitions that correspond to the new centroids. This process continues until we satisfy some stopping criterion. This procedure, despite being relatively simple, is quite powerful and is widely used in both machine learning and data compression. While these are the essentials of the process, vari-

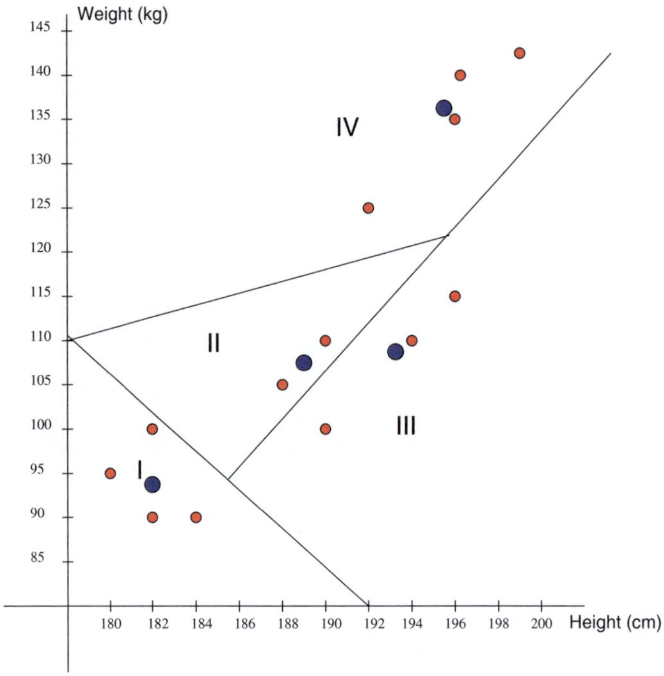

Fig. 5.3 A possible four-level classifier

ations of the basic procedure exist that address the computation of centroids, initialization of the algorithm, and the stopping criterion. Let's take a look at some of these.

The centroid calculation in the *k-means algorithm* is progressive. The calculation of the centroid is updated each time a new feature vector is assigned to a region. In the case of the *Generalized Lloyd Algorithm (GLA)*, we first assign all feature vectors to their closest centroid. The new centroids are then computed. Therefore, instead of progressively updating the centroid, the new centroid is calculated after all the feature vectors have been assigned. This makes the process more stable, so for the remainder of this chapter, we will follow the GLA approach.

The Generalized Lloyd Algorithm, as the name implies, is a generalization of an algorithm developed by Stuart Lloyd at Bell Labs [12] for the optimum partitioning of the real number line used in the design of scalar quantizers. (The work was performed in the late fifties but, for various reasons, was only published in 1982.) The extension of this algorithm to multiple dimensions and for use with a *training set* of vectors was done independently by Edward E. Hilbert at the NASA Jet Propulsion Laboratories in Pasadena [13] and by Linde, Buzo, and Gray [14]. The GLA requires the existence of a distance measure $d(X, Y)$, which is usually taken to be the mean-squared distance

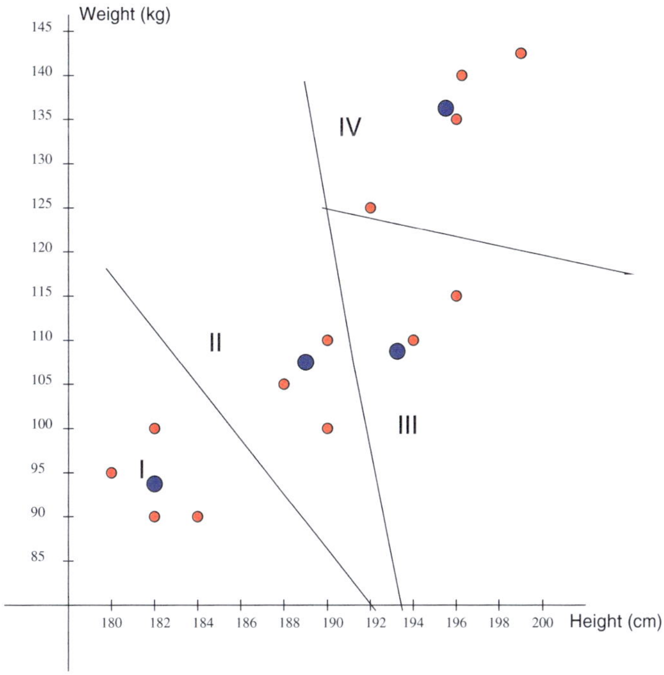

Fig. 5.4 A possible four-level classifier

$$d(X, Y) = \sum_{i=1}^{k} (x_i - y_i)^2$$

where $X = [x_1, x_2, \ldots, x_N]$ and $Y = [y_i, y_2, \ldots, y_N]$.

Suppose we are given N feature vectors $\{X_n\}_{n=1}^N$ and we want to partition them into M classes. We can summarize the GLA design process as follows [15]:

1. Select an initial set of class representatives $\left\{C_i^{(0)}\right\}_{i=1}^M$ where M is the number of desired classes. Set the counter k to zero and the average distortion $D^{(0)} = 0$. Select a stopping threshold ϵ.
2. Assign the N feature vectors to M classes. A vector X_n is assigned to class i if $d(X_n, C_i) < d(X_n, C_j)$ for all the $j \neq i$. Therefore, the set of vectors assigned to class i at iteration k is given by
$$\{X_n : d(X_n, C_i) < d(X_n, C_j)\ \forall j \neq i\} \qquad i = 1, 2, \ldots, M.$$
3. Compute the average distance $D^{(k)}$ between the training vectors and the representative reconstruction value.
4. If $\frac{(D^{(k)} - D^{(k-1)})}{D^{(k)}} < \epsilon$, stop; otherwise, continue.

Table 5.1 Set of height-weight data

Height (cm)	Weight (kg)
188	105
190	100
180	95
184	90
194	110
196	115
196	140
194	135
192	125
190	110
182	90
182	100
187	115
193	135
194	130
198	105
199	142
196	100

5. Increment k. Find new class representatives $\left\{C_i^{(k)}\right\}_{i=1}^{M}$ as the average value of the elements assigned to each class. Go to Step 2.

What happens if we have class representatives to whom no vectors are assigned? In this case, we could take the class with the largest number of vectors and split it, and generate two new class representatives.

To see how this algorithm functions, consider the following example in two dimensions.

Example: Suppose our training set consists of the height and weight values shown in Table 5.1. Let's initialize the centroids to the following values. (For ease of presentation, we will always round the coordinates of the centroids to the nearest integer.)

Height (cm)	Weight (kg)
180	120
200	140
190	130
198	120

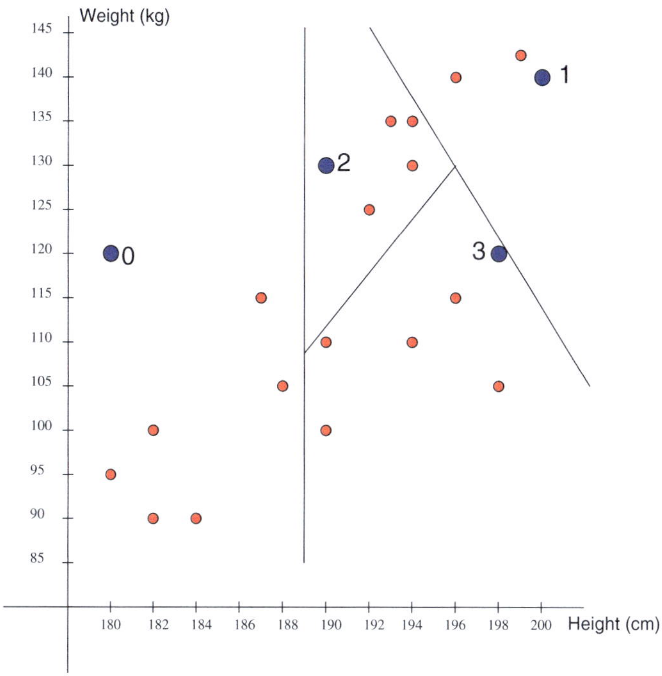

Fig. 5.5 Initial state of the classifier

The initial state of the classifier is shown in Fig. 5.5, where we have numbered the classes as 0, 1, 2, and 3. The feature vectors are shown in red, and the initial centroids are shown in blue. We have also drawn in the nearest neighbor partitions.

We can see the feature vector (188, 105) is assigned to Class 0, as have been the feature vectors (180, 95), (184, 90), (182, 90), (182, 100), and 187, 115). The feature vectors (196, 140) and 199, 142) have been assigned to Class 1. The feature vectors (194, 135), (192, 125), (193, 135), and (194, 30) have been assigned to Class 2. Finally, the feature vectors (190, 100), (194, 110), (196, 115), (190, 110), and (196, 100) have been assigned to Class 3. We now take all the vectors in each class and recompute the centroids. When we do that, the new centroids (to the nearest integer) become:

184	99
198	141
193	131
194	107

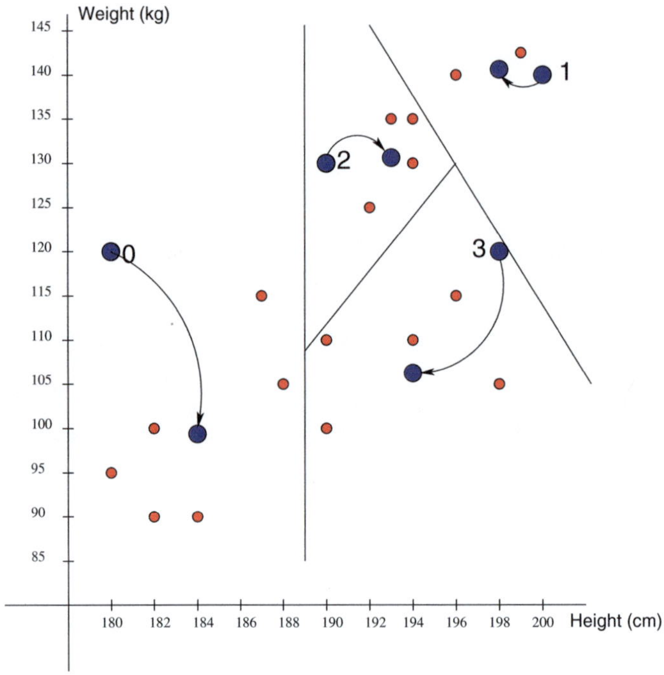

Fig. 5.6 The movement of the centroids after the first iteration

We can see from Fig. 5.6 how the centroids have moved. It is clear that they now better represent the feature vectors that fall into their respective classes.

It is also clear that the nearest neighbor partitions, and hence the assignments of the feature vectors to the centroids, no longer accurately represent the new situation. We therefore reassign the feature vectors to the new centroids. The new partitions along with the new centroids are shown in Fig. 5.7. We can see that the feature vectors (188,105) and (187,115) have moved from Class 0 to Class 3, while the feature vector (190,100) has moved from Class 3 to Class 0. Therefore, we need to update the centroids for Class 0 and Class 3. When we do that, we get the following centroids for the various classes:

184	95
198	141
193	131
193	109

Notice these are very small variations, and when we adjust the boundaries accordingly, we find that none of the feature vectors change their class, and we have reached a final state.

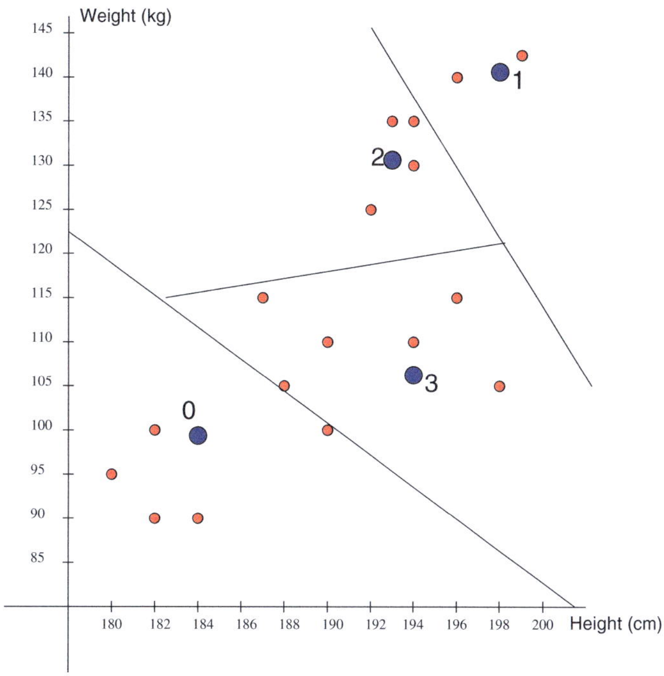

Fig. 5.7 Reassigning the feature vectors

In this particular example, we initialized the algorithm with a random selection of centroids. If we had picked a different set of centroids, we could have ended up with a different final configuration. For example, if we initialized the algorithm with the following set of values:

188	105
190	100
180	95
184	90

we obtain the clustering shown in Fig. 5.8. This is quite different from the result we obtained earlier.

It would then make sense to find a way to initialize the algorithm more systematically. E.E. Hilbert, in his work, picks the initial centroids from the set of feature vectors. Again, picking different feature vectors to start with could result in different designs. However, this approach is more robust than simply randomly selecting numbers for the centroids. Linde, Buzo, and Gray proposed a more systematic approach in their work. They called their approach a *splitting technique*. In this approach, we begin with a single class. The

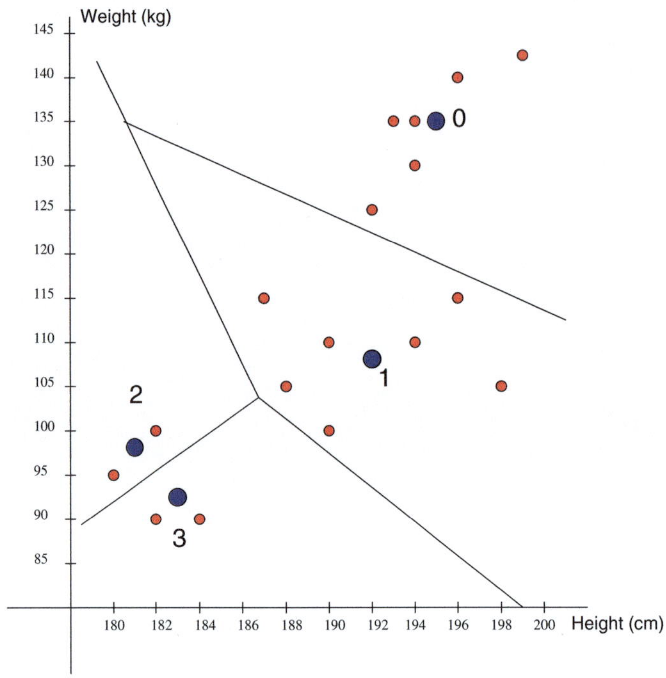

Fig. 5.8 The results using a different set of initial conditions

representative of this single class would be the average value of all the feature vectors. We then use this vector and a vector obtained by adding some fixed vector ϵ to it as the initial vectors for designing a two-class classifier. We could also pick a random value for ϵ, which would still be more robust than the other options. Once the algorithm converges to a set of values for the two-class classifier, the two centroids along with their perturbations are used as the initial vectors for a four-class classifier. We continue in this manner until we reach the desired number of classes. At first sight, it seems that by using this method, we can only obtain classifiers for which the number of classes is a power of two. However, this is not the case. Suppose we wanted a nine-class classifier. In this case, we would follow the procedure described above to obtain an eight-class classifier. We would then use the final values for the eight-class classifier and another vector obtained by perturbing the centroid of the class with the largest number of entries to get the ninth initial vector, and proceed with the same algorithm

Example: Let's revisit the previous example. This time, instead of using the initial codewords used in the example, we will use the splitting technique. For the perturbations, we will use a fixed vector $\epsilon = (10, 10)$. The perturbation vector is usually selected randomly; however, for purposes of explanation, it is more useful to use a fixed perturbation vector.

Table 5.2 Progression of codebooks using splitting

One-level		
	191	113
Two level		
	195	135
	189	103
Four-level		
	198	141
	193	131
	193	111
	186	97

We begin with a single-level classifier. The representative value of this single class is simply the average value of the training set. The progression of codebooks is shown in Table 5.2.

You can see the progression of codebooks in Fig. 5.9. In the figure, the centroid for the one-level classifier is black. It is labeled **1**, the centroids for the two-level classifier are yellow and labeled **2**, and the centroids for the four-level classifier are green and labeled **4**.

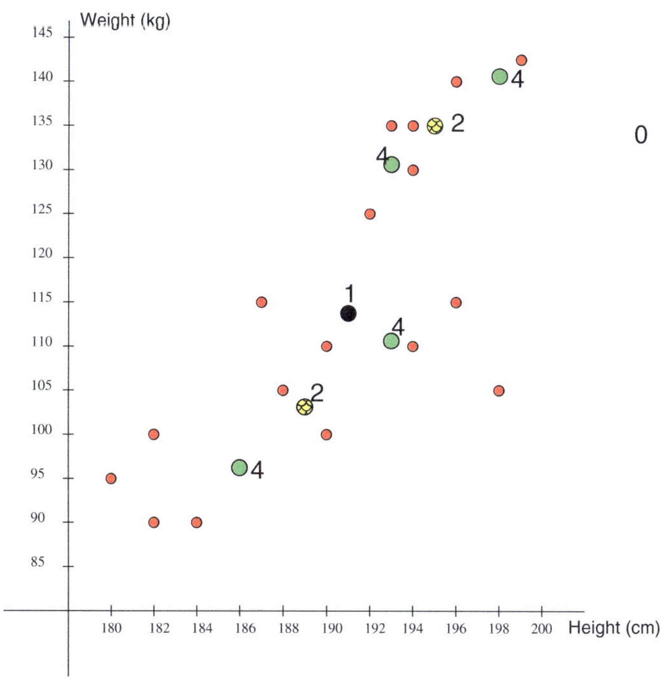

Fig. 5.9 Progression of classifiers

5.2 Support Vector Machines

> The Navy revealed the embryo of an electronic computer today that it expects will be able to walk, talk, see, write, reproduce itself and be conscious of its existence.
>
> NYT July 8, 1958

A support vector machine is a classifier that assigns feature vectors to different classes. Its roots lie in the 1958 discovery by Frank Rosenblatt of the perceptron algorithm [16]. This is the algorithm referred to in the quote above. Of course, the New York Times was wrong. Still, the perceptron algorithm is the precursor for other things the New York Times would later hype just as extravagantly, such as support vector machines, neural networks, and, nowadays, deep learning. A Support Vector Machine, while it cannot "walk, talk, see, write, reproduce itself and be conscious of its existence," is a handy binary classifier and is widely used in biomedical classification tasks. Its early incarnation was the Generalized Portrait Algorithm developed by Vladimir Vapnik and his colleagues at Moscow State University in the early 1960s [17–19]. Vapnik and colleagues [20, 21] further developed the original version into what we have today.

Given a particular binary classification task, such as determining whether a cell is cancerous or not, or whether an EKG indicates a disease, or whether an EEG corresponds to a seizure, we collect features that we hope will allow us to discriminate between these conditions. If we can assign numerical values to these features, we can generate a set of *feature vectors* $\mathbf{x}_i$ whose components are the numerical values associated with the features. Each of these feature vectors can be thought of as a point in the feature space. What we want to do is partition this feature space so that feature vectors belonging to different classes fall into distinct partitions. In the case of a linear classifier, the partition is a hyperplane that separates the feature vectors of the different classes. If we are given a *training set* consisting of feature vectors labeled with the class they belong to, we can build a classifier that works best for this training set of feature vectors. This approach for building a classifier, one in which we have a labeled training set, is called a *supervised learning* approach. Once the classifier has been built, we can use it to classify novel feature vectors. Generally, in the supervised learning approach, we are given a large set of feature vectors, which we partition into two sets. We refer to one set as the training set and use it to design the classifier. The remaining feature vectors, collectively referred to as the test set, are used to evaluate the classifier's accuracy. The Support Vector Machine is a binary classifier built using a supervised learning approach.

Consider the two-dimensional example in Fig. 5.10. In this case, there are two classes, denoted by asterisks and triangles, each represented by two features, *feature 1* and *feature 2*. The hyperplane dividing the two classes is simply a line. The line partitions the two groups exactly. Problems in which the different classes can be separated by a line (or hyperplane) are called *linearly separable* problems. In our development, we will initially look at linearly separable problems. We will later extend the solution to problems that don't fall into this class.

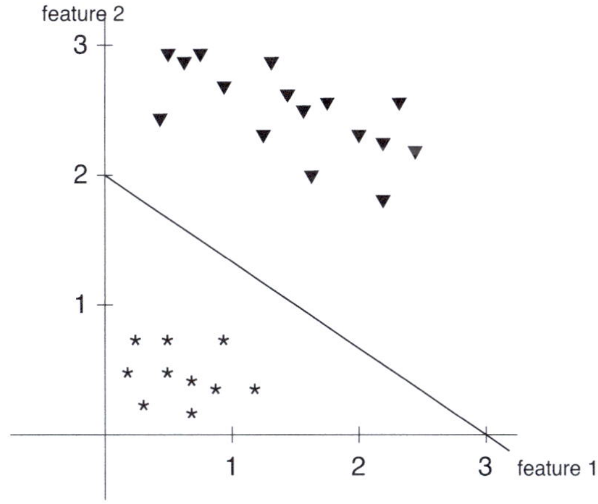

Fig. 5.10 An example of two classes separated by a line—a hyperplane in two dimensions

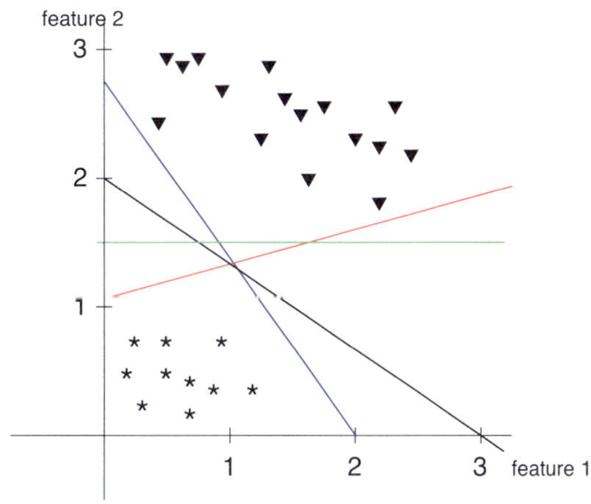

Fig. 5.11 A few of the many lines that can partition the feature space

The line shown in Fig. 5.10 is not the only line that partitions the two groups—there are infinitely many such lines; a few are shown in Fig. 5.11. So, how do we pick the "best" line? The answer given by Corinna Cortes and Vladimir Vapnik [21] was what they called the Support Vector Network, more popularly known as the Support Vector Machine.

Let's pick two parallel lines, place them between the two classes, and push them apart as far as possible so that there are no feature vectors between the parallel lines, as shown in Fig. 5.12. The distance m between these lines is called the *margin*. The partition for a Support Vector Machine is parallel to and between the two lines that provide the maximum margin. The features that fall on these lines are called the *support vectors*. In a sense, the support vectors are the most important training vectors, as they determine the partition, hence the name. Once we have determined this partition, we can use it to assign any future data points to one of the two classes depending on which side of the partition the feature vector corresponding to the data falls. For the two-dimensional case, we could plot our training set

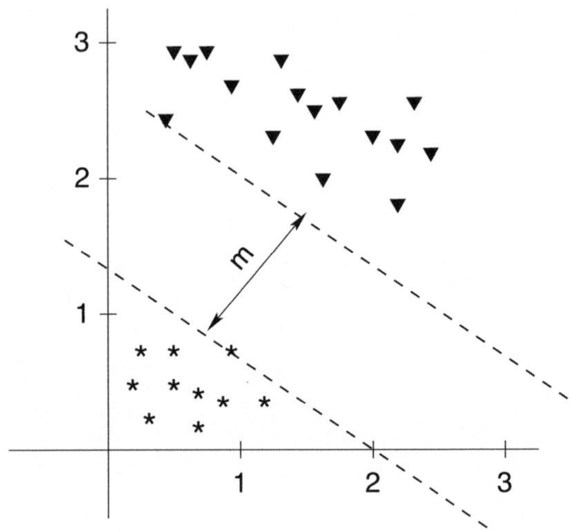

Fig. 5.12 The vector **w** is normal to the separating hyperplane

and obtain our partition by visual observation. This is not possible for higher-dimensional data. To determine how to get the Support Vector Machines for higher dimensions, we need some mathematical tools. This is what we will develop next.

5.2.1 Mathematical Development

Let's start with the equation for a line

$$y = ax - b$$

where a is the slope of the line and b is the intercept with the y-axis. For the dividing line in our example $a = -\frac{2}{3}$ and $b = -2$. For the general equation of the line, bring everything over to the left-hand side

$$-ax + y + b = 0$$

Note that x and y are the coordinates of points on the line. Let's refer to this point as a vector

$$\mathbf{x} = [x \;\; y]$$

and define the vector,

$$\mathbf{w} = [-a \;\; 1]$$

then we can rewrite the equation of the line as

$$\mathbf{w} \cdot \mathbf{x} + b = 0$$

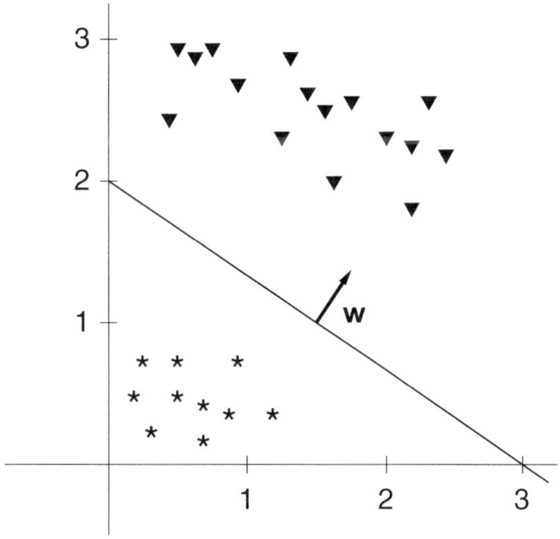

Fig. 5.13 The vector **w** is normal to the separating hyperplane

What is **w** here? We can determine that if we slide the line down so that it passes through the origin. This can be accomplished by setting the intercept b to zero.

$$\mathbf{w} \cdot \mathbf{x} = 0$$

That is, the inner product of the vector **w** with the vector corresponding to any point on the line is zero. This is true if **w** is perpendicular, or normal, to the line (Fig. 5.13). Generalizing to higher dimensions, the equation $\mathbf{w} \cdot \mathbf{x} + b = 0$ is the equation of a hyperplane with *bias* b.

Until now, we have taken **x** as a point on the line. Let's see what happens when we replace this with a point from each of the classes shown. Let's pick (0.5,2.5) from the triangle class and (0.5, 0.8) from the asterisk class. Plugging these values for **x** we get a value of 0.833 for $\mathbf{x} = [0.5 \ \ 2.5]$ and -0.866 for $\mathbf{x} = [0.5 \ \ 0.8]$. In fact, if we pick any element from the triangle class, the hyperplane equation will evaluate to a positive value, and the vectors from the asterisk class will result in a negative value. Let's assign labels to each class. We will assign a label of 1 to the triangle class and a label of -1 to the asterisk class. We can automatically get the label y_i of a point $\mathbf{x}_i$ by using

$$y_i = sgn(\mathbf{w} \cdot \mathbf{x} + b)$$

where $sgn(x)$ is the signum function

$$sgn(x) = \begin{cases} 1 & x > 0 \\ -1 & x < 0 \end{cases}$$

The question for us is, given a set of feature vectors belonging to two classes, how do we determine a line that separates these classes? Or, in terms of our formulation, given a set of feature vectors, how do we find $\mathbf{w}$ and b such that the equation shown above generates the correct labels for each of the vectors? Rather than dealing with $\mathbf{w}$ and b separately, we can also augment the vector $\mathbf{w}$ to include b. Our new $\mathbf{w}$ would then be defined as

$$\mathbf{w} = [b \quad -a \quad 1]$$

If we now define a new $\mathbf{x}$ as

$$\mathbf{x} = [1 \quad x \quad y]$$

we can write the line equation as

$$\mathbf{w} \cdot \mathbf{x} = 0$$

In this case, our classification function would become

$$f(\mathbf{x}_i) = sgn(\mathbf{w} \cdot \mathbf{x}_i)$$

Let's stick with our original formulation of the classification function and find the best dividing hyperplane—line, in this case, between two classes. Let's define two parallel lines as shown in Fig. 5.12. What we would like to do is to spread these parallel lines as far as possible while keeping the region between them empty. We can then draw our separating line as a line parallel and equidistant from these two lines. As we mentioned earlier, the distance between the two lines, m, is called the *margin*. What we would like to do is maximize the value of m. To do that, we need first to find an expression for m.

Because the three lines are parallel and because the central line is described by

$$\mathbf{w}^T \mathbf{x} + b = 0$$

we can write the equations for the other two parallel lines as

$$\mathbf{w}^T \mathbf{x} + b = \delta$$

for the upper line and

$$\mathbf{w}^T \mathbf{x} + b = -\delta$$

for the lower line. Remember, they are equidistant from the central line. We can now scale everything so that $\delta = 1$. This means that for all points $\mathbf{x}_i$ in the triangle cluster

$$\mathbf{w}^T \mathbf{x}_i + b \geq 1$$

and for all points in the asterisk cluster

$$\mathbf{w}^T \mathbf{x}_i + b \leq -1$$

If as before

$$y_i = sgn(\mathbf{w} \cdot \mathbf{x}_i + b)$$

then

$$y_i(\mathbf{w}^T\mathbf{x}_i + b) \geq 1$$

for all the points.

Pick a point $\mathbf{x}_1$ on the lower line and let $\mathbf{x}_2$ be a point on the upper line such that the line connecting the two is perpendicular to all three lines (see Fig. 5.14). The vector $\mathbf{w}$ is perpendicular to all three lines, so we can get the unit vector perpendicular to all three lines as $\mathbf{w}/|\mathbf{w}|$. The distance between $\mathbf{x}_1$ and $\mathbf{x}_2$ is m so we can write $\mathbf{x}_2$ in terms of $\mathbf{x}_1$ as

$$\mathbf{x}_2 = \mathbf{x}_1 + m\frac{\mathbf{w}}{|\mathbf{w}|}$$

Recall,

$$\mathbf{w}^T\mathbf{x}_2 + b = 1$$

Substituting the expression for $\mathbf{x}_2$ from above this means that

$$\mathbf{w}^T\left(\mathbf{x}_1 + m\frac{\mathbf{w}}{|\mathbf{w}|}\right) + b = 1$$

or

$$\mathbf{w}^T\mathbf{x}_1 + b + m\frac{\mathbf{w}^T\mathbf{w}}{|\mathbf{w}|} = 1$$

but because $\mathbf{x}_1$ is a point on the lower line

$$\mathbf{w}^T\mathbf{x}_1 + b = -1$$

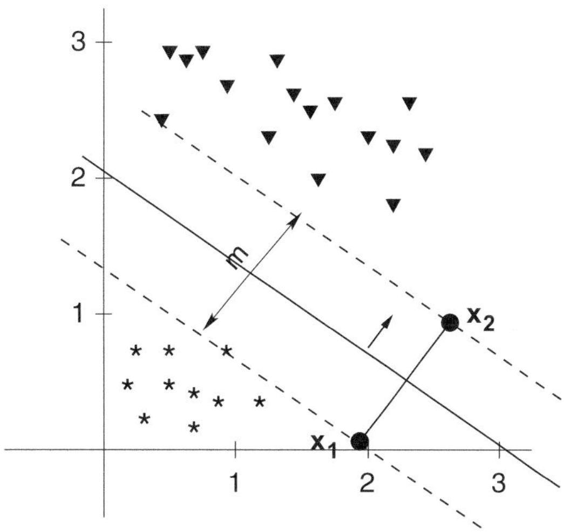

Fig. 5.14 The vector $\mathbf{w}$ is normal to the separating hyperplane

Substituting this, we have

$$-1 + m\frac{\mathbf{w}^T\mathbf{w}}{|\mathbf{w}|} = 1$$

or

$$m = 2\frac{|\mathbf{w}|}{\mathbf{w}^T\mathbf{w}}$$

and we have our expression for m. Noting that $\mathbf{w}^T\mathbf{w} = |\mathbf{w}|^2$ we can simplify this to

$$m = \frac{2}{|\mathbf{w}|}$$

We want to maximize m. However, this is equivalent to minimizing $|\mathbf{w}|$. Could we then pick $\mathbf{w} = 0$? No, because we still have to satisfy the requirement that

$$y_i(\mathbf{w}^T\mathbf{x}_i + b) \geq 1$$

for all the points. Also minimizing $|\mathbf{w}|$ is the same as minimizing $|\mathbf{w}^2|$. Minimizing $|\mathbf{w}|^2$, which is the same as $\mathbf{w}^T\mathbf{w}$, is mathematically more convenient. The optimization problem to find the best classifier partition becomes

$$\min_{\mathbf{w},b} \frac{|\mathbf{w}|^2}{2} \text{ subject to } y_i(\mathbf{w}^T\mathbf{x}_i + b) \geq 1 \ \forall\, \mathbf{x}_i$$

or,

$$\min_{\mathbf{w},b} \frac{\mathbf{w}^T\mathbf{w}}{2} \text{ subject to } y_i(\mathbf{w}^T\mathbf{x}_i + b) - 1 \geq 0 \ \forall\, \mathbf{x}_i$$

Note that, although all our illustrations are for a two-dimensional problem, the formulation is general and can be applied to arbitrary dimensions. This is a constrained optimization problem, and several well-known methods, such as the Lagrangian multiplier method, can be used to solve for $\mathbf{w}$ and b.

In the Lagrangian formulation, we set up the function to be minimized

$$L = \frac{1}{2}\mathbf{w}^T\mathbf{w} - \sum_i \lambda_i[y_i(\mathbf{w}^T\mathbf{x}_i + b) - 1]$$

We will leave the actual solution to numerical functions available in whatever language we are using to implement our particular problem, but let's go through some of the math so that we can see how the L is related to the training set of feature vectors. We take the derivative of L with respect to $\mathbf{w}$ and b.

$$\frac{\delta L}{\delta \mathbf{w}} = \mathbf{w} - \sum_i \lambda_i y_i \mathbf{x}_i = 0$$

$$\frac{\delta L}{\delta b} = = -\sum_i \lambda_i y_i = 0 \Rightarrow \sum_i \lambda_i y_i = 0$$

From the first equation we can see that the vector $\mathbf{w}$ can be found as a linear combination of the feature vectors.

$$\mathbf{w} = \sum_i \lambda_i y_i \mathbf{x}_i$$

Let's plug this into the expression for L

$$L = \frac{1}{2}\left(\sum_i \lambda_i y_i \mathbf{x}_i\right)^T \left(\sum_j \lambda_j y_j \mathbf{x}_j\right) - \sum_i \lambda_{[} y_i \mathbf{x}_i^T \left(\sum_j \lambda_j y_j \mathbf{x}_j\right) - \sum_i \lambda_i y_i b + \sum_i \lambda_i$$

or

$$L = \frac{1}{2}\left(\sum_i \lambda_i y_i \mathbf{x}_i\right)^T \left(\sum_j \lambda_j y_j \mathbf{x}_j\right) - \left(\sum_i \lambda_i y_i \mathbf{x}_i^T\right)\left(\sum_j \lambda_j y_j \mathbf{x}_j\right) - b\sum_i \lambda_i y_i + \sum_i \lambda_i$$

where we simply put parentheses around the first sum in the second term and removed the constant b from the summation in the third term. The third term then becomes zero, and we have

$$L = -\frac{1}{2}\left(\sum_i \lambda_i y_i \mathbf{x}_i\right)^T \left(\sum_j \lambda_j y_j \mathbf{x}_j\right) + \sum_i \lambda_i$$

or

$$L - \sum_i \lambda_i - \frac{1}{2}\sum_i \sum_j \lambda_i \lambda_j y_i y_j \mathbf{x}_l^T \mathbf{x}_j$$

The dependence of L, the quantity we want to minimize to find $\mathbf{w}$ and b, on the training set vectors is through the inner products of the training set vectors. Once we have the vector $\mathbf{w}$ and the bias b, classification is simply a matter of finding the inner product between the data point we are trying to classify and $\mathbf{w}$ and adding the bias to it. If the result is a value greater than one, the data point belongs to one class. If the result is a value less than negative one, the data point belongs to the other class. Again, for the computations we have to perform on the vectors, all we need are inner products. This will become important when we look at the general case of Support Vector Machines.

MATLAB Example The support vector machine is a widely used classifier. As such, many variations of the classifier are implemented in MATLAB, Python, and other languages. In MATLAB, one implementation is the *fitcsvm* function. Let's use this function to generate a Support Vector Machine. First, let's generate ten members of two data classes similar to the data we have been using in our illustrations.

```
 clear all
n = 20; % Number of feature vectors
x1 = [0.5 0.5] + 0.5*randn(floor(n/2),2); % Class 1
x2 = [2.5 2.5] + 0.5*randn(n-floor(n/2),2); % Class 2
figure
```

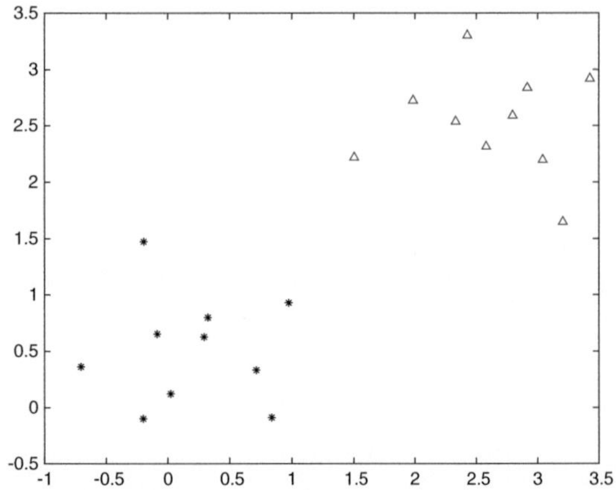

Fig. 5.15 Two classes of feature vectors generated for the MATLAB example

```
plot(x1(:,1),x1(:,2),'k*')
hold on
plot(x2(:,1),x2(:,2),'b^')
```

This results in the feature vectors shown in Fig. 5.15.

We then put the feature vectors, generate the labels showing the class members, and then use the *fitcsvm* function to generate the Support Vector Machine model.

```
x = [x1;x2];
labels = ones(n,1); % labels for the first class
labels(floor(n/2)+1:n)= -1;% labels for the second class
model = fitcsvm(x,labels,'KernelFunction','linear')
figure
plot(x1(:,1),x1(:,2),'k*'); % plot Class 1
hold on
plot(x2(:,1),x2(:,2),'b^'); % plot Class 2
sv = model.SupportVectors; %obtain the support vectors
hold on
scatter(sv(:,1),sv(:,2),100,'b','filled'); % plot support vectors
legend('Class 1','Class 2','Support Vectors');
```

The *model* structure contains, among other things, the support vectors. These are the feature vectors that provide support for the upper and lower margins. For this example, the support vectors are shown as blue-filled circles in Fig. 5.16. Using the support vectors we can draw in the margins and the boundary line as shown in Fig. 5.17.

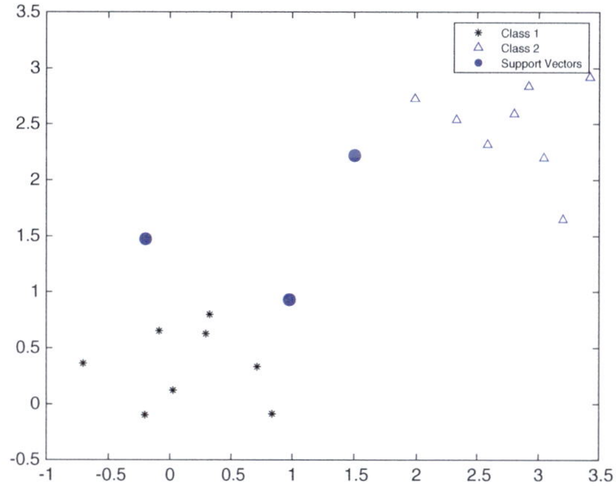

Fig. 5.16 The support vectors for the MATLAB example

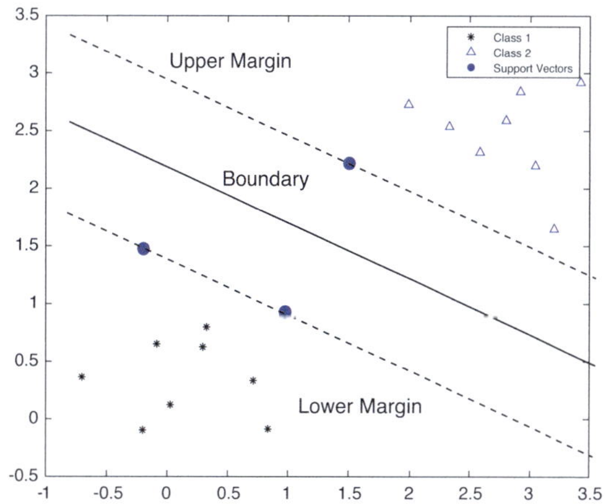

Fig. 5.17 The classifier for the MATLAB example

5.2.2 Soft Margin Extension

The example that we picked was a very easy one in that the classes could be completely partitioned using a line. The margin obtained in such a case is called a *hard margin*. What if we had the situation shown in Fig. 5.18 in which we could get a much larger margin if we didn't insist on classifying everything correctly—we allowed some slack?

In this situation, if we used the original separating line, we would misclassify one of the asterisks. However, we would end up with a larger margin for the remaining inputs. If we moved the separating line to accommodate the outlier, we would correctly classify all the training data, but end up with a smaller margin. As the dividing boundary goes through the middle of the margin, if the test data looked like the majority of the training data, this would mean that we would have a higher probability of misclassifying some of the triangle data, as the triangle cluster would be much closer to the dividing partition. What we have is a

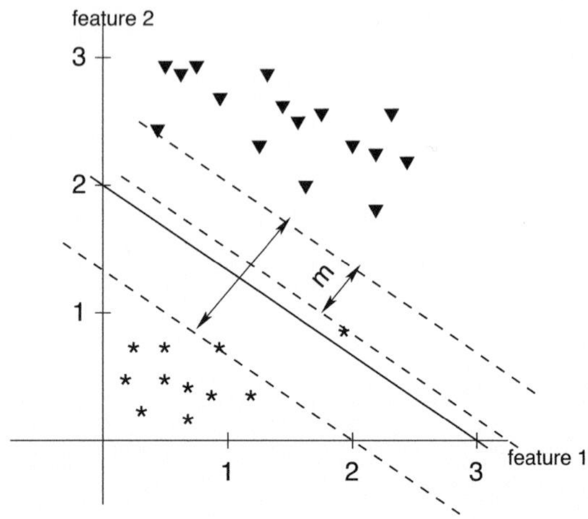

Fig. 5.18 The difference between soft and hard margins

tradeoff. This tradeoff is formally incorporated into the Support Vector Machine formulation by changing the decision rule from

$$\mathbf{w}^T\mathbf{x}_i + b \geq 1 \Rightarrow y_i = 1$$
$$\mathbf{w}^T\mathbf{x}_i + b \leq -1 \Rightarrow y_i = -1$$

to

$$\mathbf{w}^T\mathbf{x}_i + b \geq 1 - \zeta_i \Rightarrow y_i = 1$$
$$\mathbf{w}^T\mathbf{x}_i + b \leq -(1 - \zeta_i) \Rightarrow y_i = -1$$

for the feature vectors that fall within the margin, with ζ_i being the distance from the margin. To make the formulation general, we can apply the decision rule above to all feature vectors in the training set and set ζ_i to zero for the feature vectors $\mathbf{x}_i$ that fall outside the margin. In other words,

$$\zeta_i = max\left(0, 1 - y_i(\mathbf{w}^T\mathbf{x}_i + b)\right)$$

The margin we obtain in this situation, where we allow some training vectors to fall within the margin or even be misclassified, is called a *soft margin*.

The larger ζ_i is, the higher the cost incurred. This is incorporated into the problem formulation by changing the quantity to be minimized from $|\mathbf{w}|^2$ to $|\mathbf{w}|^2 + C\sum_i \zeta_i$, where C is a positive constant. If C is large, the cost of letting feature vectors fall within the margin becomes high, and we will generally end up with a narrow margin. If C is small, then the cost of accepting some mistakes in the training set is less, and we can expand the margin at the expense of making a few errors. We also expand the definition of support vectors to include the feature vectors that fall inside the margin.

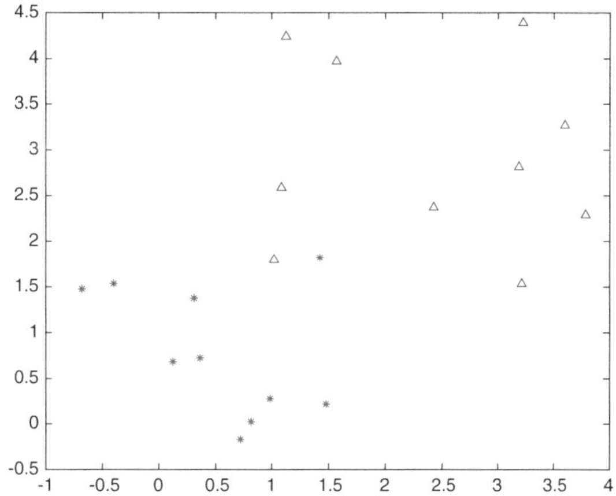

Fig. 5.19 Feature vectors that are not linearly separable

MATLAB Example We can generate a dataset for this more complicated example by just changing the variance of the random vectors in our earlier example.

```
clear all
n = 20; % Number of feature vectors
x1 = [0.5 0.5] + 1.5*randn(floor(n/2),2); % Class 1
x2 = [2.5 2.5] + 1.5*randn(n-floor(n/2),2); % Class 2
figure
plot(x1(:,1),x1(:,2),'r*')
hold on
plot(x2(:,1),x2(:,2),'b^')
```

The result is the set of feature vectors shown in Fig. 5.19.

Using the same *fitcsvm* function, we get the partition and the margins shown in Fig. 5.20. The support feature vectors are circled in the figure. Notice that the support vectors include both the feature vectors defining the margins and the feature vectors that fall inside the margin or are misclassified.

5.2.3 The Kernel Trick

What if we have a situation where the feature vectors are distributed in a way that a hyperplane cannot separate them? Consider the situation in Fig. 5.21. In this case, there is no line that we can draw to distinguish between the two clusters.

However, suppose we were allowed to draw a circle instead of a straight line as the boundary. In that case, we can see that a circular boundary would permit us to partition the feature space into two regions, one within the circle and one outside the circle. But this would take us away from the comfort of drawing a straight line to partition the space. Not

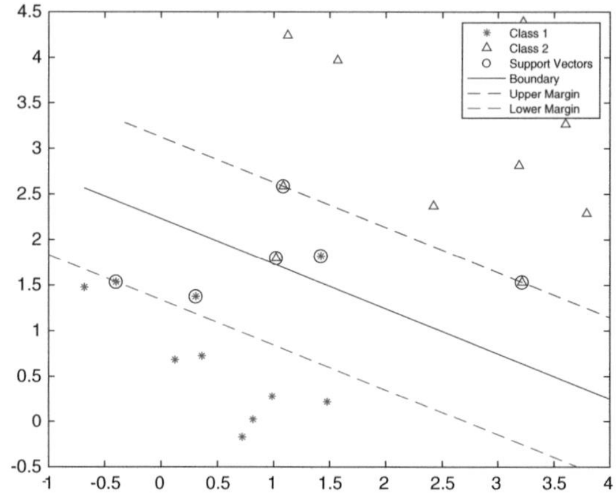

Fig. 5.20 Support Vector Machine for the case where the feature vectors are not linearly separable. The support vectors are circled

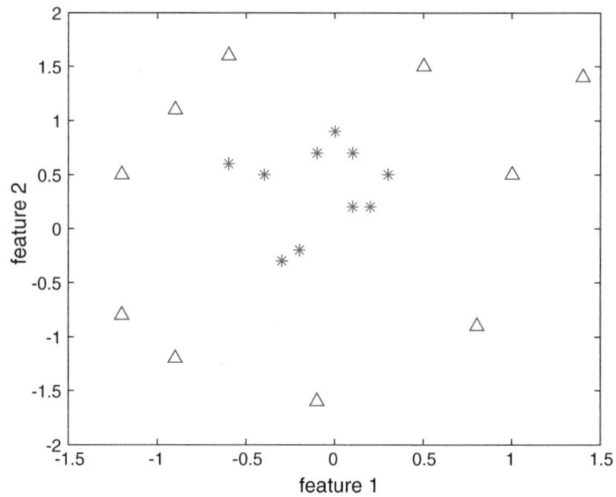

Fig. 5.21 A situation where a hyperplane cannot separate the two classes

only that, but remember that we can classify a novel feature vector **u** by simply evaluating $\mathbf{w}^T\mathbf{u} + b$ and checking the sign of the result. If the result is positive, the new vector belongs to one class; if it is negative, then the new vector belongs to the other class. In this particular case, we can preserve these advantages if we treat each element of the feature vector as a coordinate in the Cartesian plane and convert it to polar coordinates. If we did that, we would get the situation shown in Fig. 5.22

Notice that after transformation, the feature vectors are linearly separable. If we now use the same MATLAB function to find the separation boundary, the margins, and the support vectors, we obtain the situation in Fig. 5.23.

Now, when we get a vector to classify, we can transform it into polar coordinates and evaluate $\mathbf{w}^T\mathbf{u} + b$ and check the sign of the result to obtain the classification.

This was a situation in which we could visually observe how the feature vectors were distributed in feature space, and a simple transformation was applied to move the feature

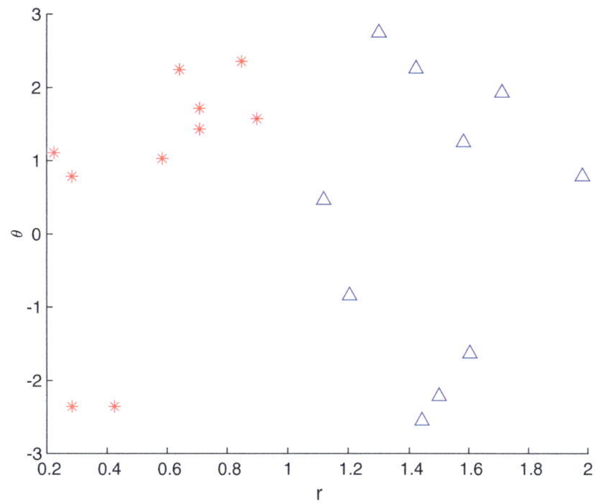

Fig. 5.22 Transforming feature vectors

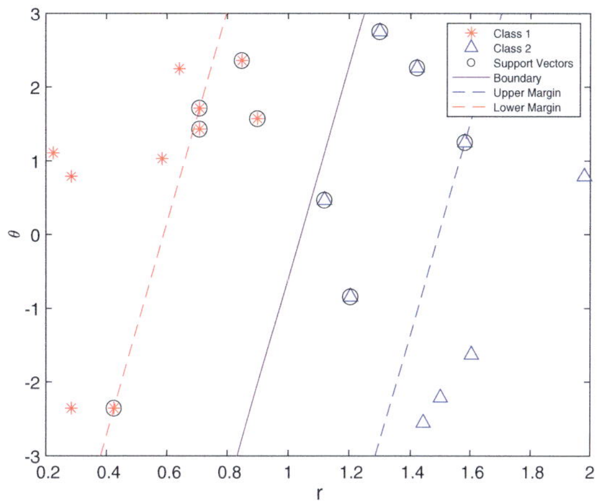

Fig. 5.23 Support Vector Machine for the transformed feature vectors

vectors into a space where they were linearly separable. Things would have been much more difficult if the feature vectors were of high dimensionality and the number of feature vectors in the training set was also large. With high dimensionality, it would be very difficult to visualize the vectors; the partition would probably not be obvious, and transforming the vectors into new spaces to evaluate their separability would be computationally expensive.

Luckily for us, Vapnik and colleagues came up with a trick, called the *kernel trick* that avoids all these problems. Let's examine the Lagrangian function we aim to minimize.

$$L = \sum_i \lambda_i - \frac{1}{2} \sum_i \sum_j \lambda_i \lambda_j y_i y_j \mathbf{x}_i^T \mathbf{x}_j$$

We can rewrite this as

$$L = \sum_i \lambda_i - \frac{1}{2} \sum_i \sum_j \lambda_i \lambda_j y_i y_j \mathbf{x}_i \cdot \mathbf{x}_j$$

to emphasize the fact that the function depends on the training set vectors only through the inner product of the training set vectors. So, if we were going to evaluate a transformation $\phi(\cdot)$, we only need the inner products $\phi(\mathbf{x}_i) \cdot \phi(\mathbf{x}_j)$. Vapnik and colleagues defined a kernel function

$$K(\mathbf{x}_i, \mathbf{x}_j) = \phi(\mathbf{x}_i) \cdot \phi(\mathbf{x}_j)$$

which could be substituted for the inner product in the Lagrangian function to evaluate a whole variety of transformations. Some common kernels are:

- Linear Kernel: This is the simplest of the Kernels, where the dot product of two vectors is simply the dot product.

  $$K(\mathbf{x}_i, \mathbf{x}_j) = \mathbf{x}_i^T \mathbf{x}_j$$

  This is the kernel we used for the hard and soft linear separation examples.
- Radial Basis Function (RBF) kernel: This is the most popular kernel and has a value that depends on the Euclidean distance between two vectors.

  $$K(\mathbf{x}_i, \mathbf{x}_j) = exp\left(-\gamma ||\mathbf{x}_i - \mathbf{x}_j||^2\right); \quad \gamma > 0$$

  If γ is replaced by σ^2, this kernel is also known as the *Gaussian Kernel.*
- Polynomial Kernel: This kernel is particularly useful when there is a discernible difference between the classes, but the boundary is wiggly. Along with the RBF kernel, it is also widely used.

  $$K(\mathbf{x}_i, \mathbf{x}_j) = \left(\mathbf{x}_i^T \mathbf{x}_j + c\right)^d$$

  where the most common value of d is two. If c is zero the kernel is sometimes called the homogeneous polynomial kernel.
- Sigmoid Kernel: The sigmoid function is widely used in neural networks. It is given by

  $$K(\mathbf{x}_i, \mathbf{x}_j) = \tanh\left(c\mathbf{x}_i^T \mathbf{x}_j + d\right)$$

If we use the polynomial kernel for the data in the previous example, we get the result shown in Fig. 5.24.

Linear kernels are less computationally expensive and, therefore, easier to implement and faster. So, wherever possible, they should probably be used. When linear kernels cannot be used, there are no hard and fast rules for which of the nonlinear kernels to use. Generally, for image processing tasks, people use polynomial kernels, while for textual analysis, RBF kernels are more commonly used. It is best to try different kernels to determine which one is most suitable for a particular application.

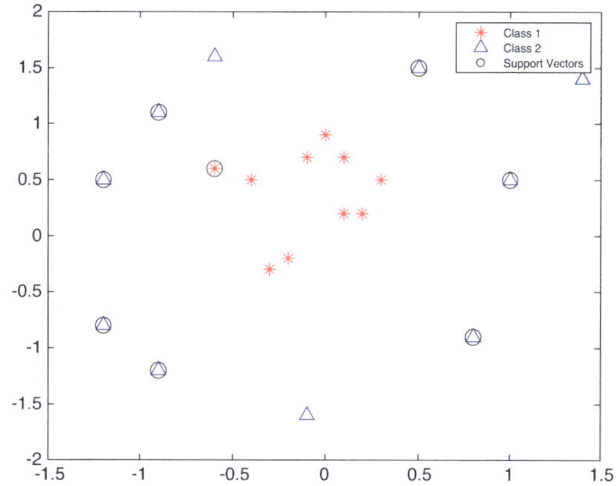

Fig. 5.24 Support Vector Machine using the polynomial kernel

5.3 Summary

Once we have a set of N features that we feel are discriminative, we can treat them as coordinates of a vector in an N-dimensional space. The problem of classification then becomes the problem of partitioning the N-dimensional space into M regions where M is the number of classes. We looked at two widely used approaches to partitioning the feature space. The Generalized Lloyd Algorithm, closely related to the k-means algorithm, partitions the space iteratively by sequentially generating partitions and class representatives, which are the centroids of the partitions. We also examined Support Vector Machines in the context of binary classification. We saw how, from an infinite number of partitions, we can select a partition that maximizes the distance between the closest members, called the margin, of the two classes. We examined the case of linearly separable clusters, which can be easily partitioned using a hyperplane. Then we described the Kernel trick, which can be used to transform a problem that is not linearly separable into one that is. There are many other classification approaches, including artificial neural nets and various regression approaches. Understanding the classification approaches presented in this chapter should facilitate understanding of other classification approaches.

Neuronal Signalling

6

The human body is a complex organization that, to function, requires continuous transfer of information between its parts. Many of the biomedical signals of interest to us are, in some form, a measurement of this process. The transfer of information through nerves and in the brain is an electrochemical process, whose basic component is neuronal transmission. Unlike the almost instantaneous transmission of an electrical signal through a wire, the transmission of a signal through a neuron is both complex and relatively slow. The transmission of a signal is initiated by a change of electrical polarization across the neuronal membrane, called the *action potential*, which then triggers a sequence of action potentials in a domino-like fashion along the axon. This sequence of action potentials is the neuronal transmission of information. In this chapter, we briefly examine the somewhat convoluted history of the discovery of neuronal transmission mechanisms and, again briefly, describe the process of neuronal transmission.

6.1 Early History

The existence of information transmission through the human body has been evident for a long time. The details of how this process works have taken a while to discover.

One of the earliest recorded attempts to understand how information transmission occurs in the body was the work of Herophilus. Born in 335 BCE in the city of Chalcedon, in the kingdom of Bithynia, now part of the municipality of Kadikoy in Istanbul, Turkey, Herophilos did most of his work in Alexandria, Egypt, during a period when dissection of human corpses was permitted. (There are stories of Herophilos taking the knife to live humans as well, but those stories are most likely not true.) The original writings of Herophilos did not survive the destruction of the Library of Alexandria in 272 CE. Still, his works were

K. Sayood, *Biomedical Signal Processing*, Synthesis Lectures on Biomedical Engineering,
https://doi.org/10.1007/978-3-032-08855-0_6

widely quoted, especially by Galen of Pergamon, one of the most influential physician-philosophers of the Roman era. Galen, in turn, was influential in the work of Ibn Sina (Avicenna), whose encyclopedic work, *The Canon of Medicine*, written in 1025 CE, was used as a medical textbook in Europe until the seventeenth century, thereby ensuring the preservation of Herophilus' works. Among other things, Herophilus studied blood flow and, importantly, when examining the brain, observed the distinction between blood vessels and nerves, which led him to propose the brain as the organ responsible for decision-making. This was a departure from Aristotle's view, who, perhaps due to an inability to differentiate between nerves and blood vessels, had proclaimed the heart as the seat of human reason. Aristotle cast a long shadow. Despite the work of Herophilus and Galen, Aristotle's view that the heart was the decision-making organ remained the dominant view for over a thousand years. The similarity between blood vessels and nerves also led to the belief that information was conveyed in the nerves through the transportation of fluids. This view also persisted for a long time. In *Treatise on Man* written in 1630, René Descartes describes two kinds of fluids, or *spirits*, vital spirits carried by arteries, and animal spirits carried by nerves. (This work was published posthumously, as Descartes got spooked by what had happened to Galileo.) The person who put the idea of animal spirits to rest was a Dutch biologist, Jan Swammerdam, who ironically was a devotee of René Descartes. In 1675, he showed that by touching an isolated frog nerve connected to a muscle with a metal plate, he could cause the muscle to contract even though there was no source for any fluid to flow into the nerve. He also demonstrated, through a rather painstaking experiment, that this process of nerve excitation and muscle contraction caused no change in volume, which would have been necessary if a fluid flow were taking place. Unfortunately for Swammerdam and the field, Swammerdam was not well-connected, and his brilliant work went almost completely unnoticed. More than a century would pass before Luigi Galvani would also touch a muscle-connected frog nerve with a metal plate to cause muscle contraction. This time, the world took notice, though not without some dissension. Galvani initially stimulated the frog nerve with a Leyden jar, leading to a muscle contraction, and argued that this meant that the nerve signal was electrical rather than fluidic. He was attacked by many who said that the electrical input was external and, therefore, was not indicative of an intrinsic electrical signal in the nerve. Galvani ditched the Leyden jar and found that a brass wire connected to the nerve would cause the frog's leg to twitch when it touched the iron tray on which the dead frog lay. The objection to this experiment came from Alessandro Volta, a professor of Physics at the University of Pavia, who said that while there was no Leyden jar to generate the electricity, the electricity was being generated by the contact between two dissimilar metals—the brass wire and the iron tray. Volta followed up his insight by experimenting with different metals, resulting in the development of the Voltaic Pile, better known now as the battery, though in French and French-influenced languages it is still called an *pile*. While Volta's invention overshadowed Galvani's work, Volta's gracious acknowledgement of Galvani's contribution still gave Galvani's work some exposure. Galvani continued his quest, finally demonstrating that he could use the nerve of a frog to cause the contraction of the exposed muscle in another frog, thus establishing the existence of electrical signals within living tissue. Galvani's work inspired Mary Shelley to write *Frankenstein* and Carlo Matteucci, at the University of Pisa, to discover that injured biological tissue would generate current, which became known as *injury current*. Matteucci also detected electric current in the heart. Matteucci's work in turn influenced the work of the German physicist, and self-proclaimed frog-faddist, Emil du Bois-Reymond [22], born in the year *Frankenstein* was published—if you needed any more connections with Galvani. Du Bois-Reymond demonstrated that a voltage difference existed between the outside and inside of a nerve membrane, which he postulated was caused by a pre-existing charge difference between the two sides of the membrane. He also demonstrated that stimulating the nerve resulted in a decrease in potential on the surface of the nerve membrane, and this decrease propagated along the nerve as a "wave of relative negativity." The work of Du Bois-Reymond delivered the coup de grace to the idea of animal spirits and set the stage for our current understanding of how neuronal information transfer works. What remained was the discovery of two things: the neuron and the membrane theory.

6.2 The Neuron

The discovery of the neuron was the work of Santiago Ramon y Cajal. Born in Navarre, Spain, in 1852, his ambition was to be an artist. His father, an anatomist, helped channel his artistic aspirations into the representation of anatomical objects and, from there, into the study of medicine. At this time, the prevailing theory about the nervous system was that it consisted of a protoplasmic network called the reticulum. According to the reticular theory, the brain was not composed of individual cells. Instead, it was a single protoplasmic network of nerve fibers rather than discrete cells. Ramon y Cajal, using a staining technique developed by one of the principal proponents of the reticular theory, the Italian physician and histopathologist Camillo Golgi, showed that this was not true; just like the rest of the body, the brain was composed of cells, or neurons. For their work, Santiago Ramon y Cajal and Camillo Golgi were awarded the Nobel Prize in Physiology or Medicine in 1906. (Golgi, unconvinced of the existence of neurons, used his Nobel lecture to throw shade on Ramon y Cajal's work.)

The neuron (see Fig. 6.1) consists of the cell body called the *soma*, the *dendrites*, and the *axon* (and *axon terminals*). The soma contains all the necessary appurtenances of a cell, including the nucleus, which contains the chromosomes, as well as mitochondria, organelles, and other cellular components. It is responsible for the metabolic functioning of the neuron. It is also the integration point for signals coming into the neuron.

The dendrites are tree-like projections of the cell, with a variable number of branches. The number of branches can range from the tens to the thousands. The dendrites are the location of inputs to the neurons. They contain channels through which ionic currents can flow into the neuron. These channels open and close in response to various chemical and electrical signals. These signals can originate from other neurons or from the environment.

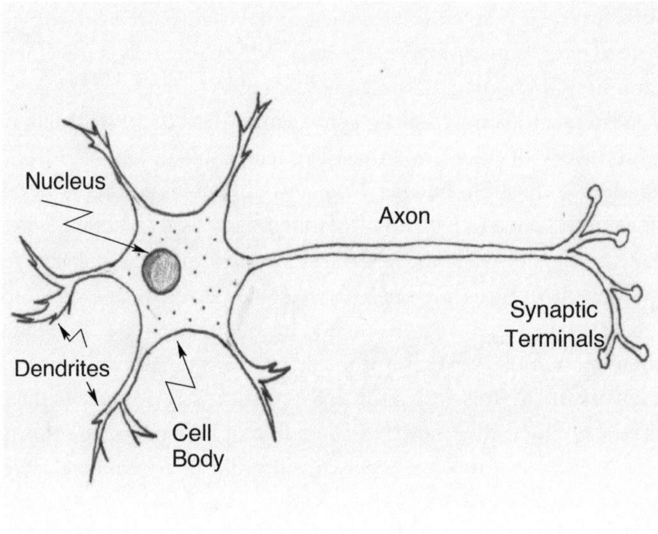

Fig. 6.1 Neuronal Cell

The axon is a long extension arising at the axon hillock and ending in a branched structure consisting of the axon terminals. In terms of signaling, we can think of the axon and its terminals as the parts of the neuron that provide the output. This output can serve as input to other neurons or to the parts of the body that perform work, such as muscle cells and glands. The transmission of a signal from one neuron to another is mostly through chemical signaling. The transmitting neuron releases various neurotransmitters, which bind to the dendrites of other neurons, thereby opening or closing ion channels. The space in which these signals are released is called a *synapse*. Ramon y Cajal first discovered the synapse. There can also be electrical signaling between closely connected cells through pathways of low resistance called *gap junctions*. This mode plays a crucial role in the signaling process that regulates the heart's function. Communication between neurons can also be influenced by the generation of electrical fields resulting from neuronal activity. A nice history and discussion of the different ways of inter-neuron communications can be found in a paper by Faber and Pareda [23]. While this chapter primarily focuses on communication between neurons, there is also communication between neurons and muscle cells and glands, and synapses are present between neurons and muscle cells or glands.

How does the neuron take input signals from its dendrites and deliver them to the axon terminal? The process of signal transmission through a neuron differs significantly from the transmission of electrical signals through a wire. In a wire, the signal flow is due to the flow of charged particles through the wire. In a nerve, the signal flow is a result of the propagation of *action potentials* through the neuron, the wave of relative negativity observed by du Bois-Reymond. To understand this process, we need to be familiar with the idea of a *membrane potential* and *ion channels*. The concept of the membrane potential was the first to be discovered.

6.3 Membrane Potential

In 1887, the German chemist Walther Nernst developed an equation for the electric potential generated across a membrane when you had unequal concentrations of a particular ion on the two sides of the membrane. During most of the nineteenth century, the general consensus was that cells did not have a membrane surrounding them. In general, this was due to the conception of the cell membrane as a rigid structure, which didn't align with the flexibility of cells. Instead, the idea was that protoplasm was hydrophobic, which led to it clumping up in the form of cells [24]. By the end of the nineteenth century, the existence of a cell membrane was becoming increasingly difficult to deny. In 1902, Julius Bernstein [25], a student of Emil du Bois-Reymond, developed an explanation for du Bois-Reymond's observation of a voltage difference between the inside and outside of a nerve membrane based on Nernst's work.

There are two types of gradients across the cellular membrane: a chemical concentration gradient and an electrical gradient. If you examine the electrolyte composition of extracellular fluids in animals, you will find relatively large concentrations of sodium ions (Na^+), chlorinc ions(Cl^-), and bicarbonatc ions ($IICO_3^-$). Thc clcctrolytc composition of intraccllular fluid, on the other hand, is primarily composed of potassium ions (K^+), magnesium ions (Mg^{++}), organic phosphates, and proteins, with a minimal amount of sodium, chlorine, and bicarbonate ions. This difference in concentration creates a concentration gradient, which the cell can utilize for various purposes. If the membrane were permeable to these ions, they would flow across the membrane in the direction of the gradient until the concentration inside and outside was the same—that is, the concentration gradient would tend to zero. To examine what actually happens, let's focus on one ion, the intracellular potassium ions. Let's suppose that initially there are enough negatively charged particles inside the cell to neutralize the positively charged potassium ions. Let's suppose the extracellular fluid is devoid of ions. If we open a pathway for only the potassium ions, due to the concentration gradient, we will see a flow of potassium ions from inside the cell to outside the cell. This will result in an increase in positive charges outside the cell and a corresponding increase in negative charges inside the cell, leading to the development of an electrical gradient across the cell membrane. This gradient would try to push the potassium ions back into the cell. The flow of potassium ions will stop when the electrical gradient balances the chemical gradient. The membrane potential at equilibrium is called the *equilibrium potential* for potassium. We can find this potential for a particular ion (such as potassium) by equating the potential due to the chemical gradient with the potential due to the electrical gradient. The potential due to the chemical gradient for potassium ions is given by

$$RT \ln \frac{[K^+]_{outside}}{[K^+]_{inside}}$$

where R is the gas constant, T is the temperature in Kelvin, and $[K^+]_{outside}$ and $[K^+]_{inside}$ are the extracellular and intracellular concentrations of the potassium ions. The electrical gradient is given by zFV_K, where z is the charge, which is +1 for potassium, F is the Faraday constant, and V_K is the potential difference across the membrane. Equating the two, we obtain the resting potential for potassium as

$$V_K = \frac{RT}{zF} \ln \frac{[K^+]_{outside}}{[K^+]_{inside}}$$

For common values of the concentrations of sodium, potassium, and chlorine, these potentials are:

$$V_{Na} = 55\text{mV}, \quad V_K = -78\text{mV}, \quad V_{Cl} = -60\text{mV}$$

The sum of these potentials is about −70 mV. This potential difference of approximately −70 mV is referred to as the *membrane resting potential.*

Based on his measurements of the resting potential, Bernstein hypothesized that at rest, the nerve membrane was only permeable to potassium, resulting in the potential difference. He also demonstrated that altering the temperature of the oil in which the nerve was submerged altered the resting potential in a linear fashion, as would be expected from Nernst's equation. He believed that the action potential was a depolarizing wave resulting from the breakdown of the membrane's permeability, allowing the membrane to become permeable to all ionic species. This would mean that the potential across the membrane would go from the resting potential of potassium (−78 mV) to zero. Bernstein's *membrane theory* became the dominant theory in the early part of the twentieth century.

6.4 The Hodgkin-Huxley Model

Bernstein's membrane theory predicted that during an action potential, the membrane would become permeable to all ionic species and the potential would drop to zero. During the 1930s Kenneth Cole and Howard Curtis conducted a series of experiments [26] in which they measured the impedance across the squid giant fiber during an action potential. As predicted by Bernstein's theory, they found a rapid decrease in impedance. This decrease in impedance was almost entirely due to a reduction in resistance from "a resting value of 1000 ohm cm.^{2} to an average of 25 ohm cm.^{2}". The capacitance of the membrane remained unchanged. They also found something unexpected.

Contrary to Bernstein's assertion, the membrane potential did not go to zero during the action potential. Instead, there was a significant overshoot. This result had also been found the previous year in crab nerves by Alan Hodgkin and Andrew Huxley, who played an essential role in the story of neuronal communication.

Alan Hodgkin and Andrew Huxley, in a series of five papers, conjectured the existence of ion channels and provided an electrical model for signal propagation via action potentials. Their work led to a Nobel Prize in Physiology or Medicine in 1963. (They shared the prize with J.C. Eccles for his work on the synapse.) Their work began with a failure. In 1938, Alan Hodgkin had spent some time at the Marine Biological Laboratory in Woods Hole working on understanding neuronal transmission along the giant squid axon. On his return to Cambridge, he continued with that work and hired Andrew Huxley, one of his more promising undergraduate students. Huxley wanted to study the viscosity of the cytoplasm in the axon (the axoplasm), so he hung up a giant squid axon and put a drop of mercury on the cut end. The idea was to monitor the progress of the mercury drop through the axoplasm. The mercury stayed put. However, while watching the axon, Hodgkin and Huxley saw a way to insert an electrode inside the axon (each credited the other for the idea), allowing them to measure the potential difference across the membrane. Previously, researchers measured potential across the membrane by placing one electrode on the membrane and the other at the cut end of the neuron, serving as a proxy for the inside of the membrane. Using this more reliable approach, the Hodgkin-Huxley method showed definitively that the action potential did not end with a zero potential across the membrane. Instead, there was a clear overshoot, which was impossible according to Bernstein's membrane theory. Their work was progressing well, but was interrupted by World War II.

After the war, Hodgkin, having survived a friendly fire incident, and Huxley restarted their research. Earlier, Bernard Katz [27] had shown that action potentials were not conducted in a salt-free solution. This followed Hodgkin's experiments before the war [28], which showed dependence of the membrane potential on the concentration of sodium outside the nerve. This indicated that the action potential depended on both the sodium and potassium permeability. As the resting potential of the membrane is close to the resting potential of potassium, this meant that the permeability of the membrane to potassium was higher than the permeability of the membrane to sodium during the resting potential. This also meant that the relative permeability changed during the action potential, which would account for the overshoot. This is in contrast to Bernstein's theory, which required the membrane to become completely permeable to all ionic species.

As the permeability changed with the changing voltage across the membrane, how does one go about calculating the permeability? The answer was a voltage clamp circuit developed by Cole and later refined by Huxley. The voltage clamp used feedback to maintain the membrane voltage at a fixed value by injecting current into the nerve. Recall that the reason you have a potential difference across the membrane is because of the movement of ions through the membrane - in other words, a current through the membrane. The current injected by the voltage clamp balances the current flowing through the membrane, keeping the voltage fixed at the desired value. In this way, we know both the voltage across the membrane and the current through it. There are two types of possible currents through the membrane: a capacitive current resulting from the capacitive nature of the membrane, and ionic currents resulting from concentration gradients. The capacitive current is proportional to the derivative of the voltage, and as the voltage is kept constant by the clamp, this current will be zero. Therefore, the current injected by the clamp is a measure of the ionic currents. If we consider permeability as the conductance of the membrane, we can determine the permeability by applying Ohm's law and dividing the current by the voltage. There is a problem, though. We know that the membrane current consists of at least two ionic currents, a Sodium current and a Potassium current, each with different permeability. So, how do we isolate the permeability of the membrane to Potassium from the permeability of the membrane to Sodium? Hodgkin and Huxley were able to isolate the sodium current and the potassium current by conducting the experiment twice: once with the membrane bathed in sodium chloride, and once with sodium chloride replaced by Choline Chloride. As Choline cannot pass through the membrane, the current in the second experiment was (mostly) due to Potassium. If we subtract this current (current due to Potassium) from the current obtained with the membrane bathed in a Sodium Chloride solution (current due to both Sodium and Potassium), we get the Sodium current. Now we can determine the conductance, or permeability, of the membrane to Sodium and Potassium for a single voltage. Hodgkin and Huxley repeated this experiment multiple times to find the conductance as a function of the voltage for both Sodium and Potassium. For Potassium, they found the best fit to their data if they assumed that the Potassium conductance was given by

$$g_K = \bar{g}_K n^4$$

where $\bar{g}_K$ is a constant and n is a dimensionless variable with a value between 0 and 1, which is a solution of the differential equation

$$\frac{dn}{dt} = \alpha_n(1-n) - \beta_n n$$

where α_n and β_n are voltage dependent rate constants. The expression for the conductance was derived by fitting an equation to the data, as this was long before the understanding of how the various ionic currents were transported across the membrane was established. Despite this, Hodgkin and Huxley in their final paper of the group of five papers [29] make a conjecture which has turned out to be quite accurate:

> *These equations may be given a physical basis if we assume that potassium ions can only cross the membrane when four similar particles occupy a certain region of the membrane.*

The sodium conductance they derive as

$$g_{Na} = \bar{g}_{Na} h m^3$$

where $\bar{g}_{Na}$ is a constant and h and m are solutions of the differential equations

$$\frac{dm}{dt} = \alpha_m(V)(1-m) - \beta_m(V)m$$

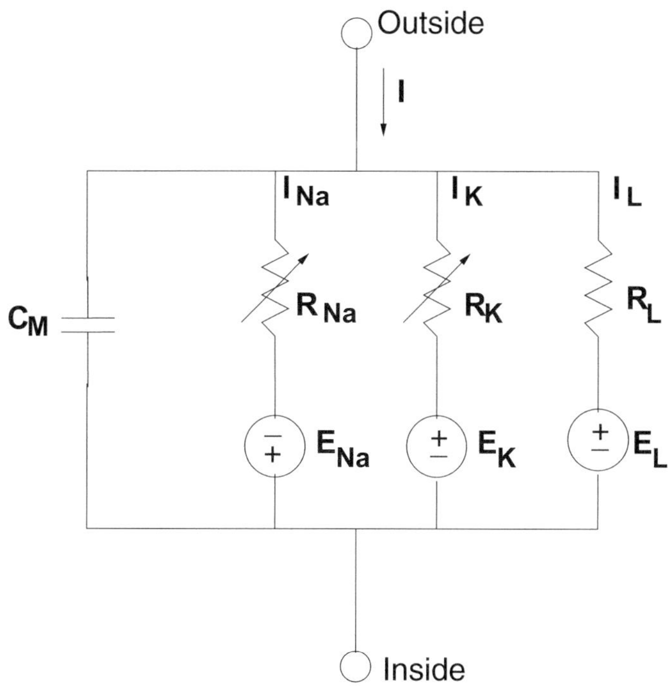

Fig. 6.2 The Hodgkin-Huxley Model for the neuronal cell membrane

$$\frac{dh}{dt} = \alpha_h(V)(1-h) - \beta_h(V)h$$

They then developed the electrical model shown in Fig. 6.2.

The membrane itself can be clearly modeled as a capacitor. Hodgkin and Huxley proposed the existence of three channels, one for Sodium ions, one for Potassium ions, and a leakage channel for Chloride and other ions. These channels are each modeled with a voltage source and a resistor—the conductance of the resistor is as described above. The voltage of the source is determined by the Nernst equation, which describes the equilibrium potential of the particular ion. We can then write the current as

$$I = C\frac{dV}{dt} + g_{Na}(V - V_{Na}) + g_K(V - V_K) + g_L(V - V_L)$$

Under equilibrium conditions, the derivative of the voltage is zero. Plugging in the expressions for the conductances, we get the membrane current during the action potential as:

$$I(t) = \bar{g}_{Na}m^3h(V - E_{Na}) + \bar{g}_K n^4(V - E_K) + \bar{g}_L(V - E_L)$$

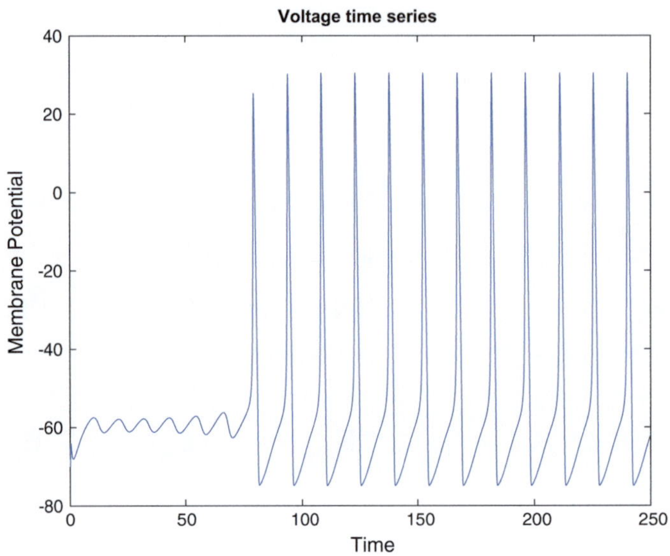

Fig. 6.3 The output from *hhrun*

There is a MATLAB function *hhrun* (developed by Rohit Chandra) [30] that you can download, which uses a forward Euler method for solving these differential equations. You can call it with

```
[V,m,h,n,t] = hhrun(I,tspan, v, mi, hi, ni,Plot)
```

where I is the current in milliamps, $tspan$ is the length of the simulation, v is the resting potential. If we call this function with the command

```
hhrun(10,250,-70,.5,.06,.5,1)
```

we get the result shown in Fig. 6.3.

6.5 Ion Channels

The model developed by Hodgkin and Huxley required that the transport across the membrane for different ionic species, in particular Potassium and Sodium, operate differently for the different ions and operate differently at different voltages. How exactly was this happening? At the time they developed their theory, there was no basis for suggesting a molecular mechanism for what was happening. The acceptance of the existence of a cell membrane itself was relatively new and contentious. One possible explanation advanced was that the

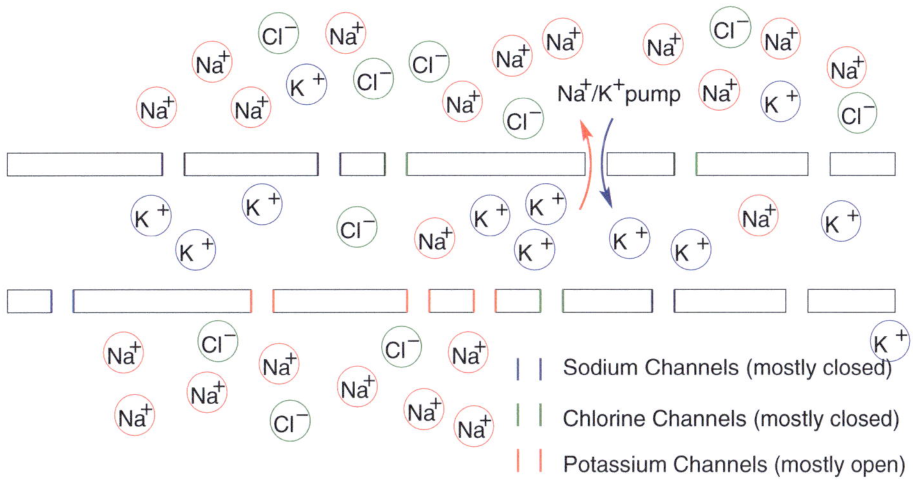

Fig. 6.4 Membrane potential

cell membrane had a dynamic structure that changed conformation during the action potential, sometimes permitting potassium to flow through and sometimes permitting sodium to flow through [31]. Based on the experimental results and the mathematical model developed by Hodgkin and Huxley, this does not seem very likely, as the changing conductance of each ion is independent of the concentration of the other ions. That would be a difficult feat if the changing conductances were a result of some global conformational change of the membrane. A more likely explanation was the existence of separate "channels" for the different ions (Fig. 6.4). In 1957, Tasaki and Hagiwara [32] found that injecting a squid axon with Tetraethylammonium Chloride led to a change in the time course of the action potential. Instead of a rapid rise and then a decay, there was a rapid rise and then a plateau. When they repeated the experiment with Sodium ions replaced by Choline, they found a decrease in the peak. Given that, according to the Hodgkin-Huxley model, the action potential requires Sodium and Potassium currents, this seemed to indicate that tetraethylammonium chloride was blocking only the Potassium current, which would support the existence of separate mechanisms for the Potassium and Sodium Currents. This was the conclusion of Bertie Hille. In a 1970 review [33] of the work in this area, Hille identifies the two questions that were the focus of attention during the 60s:

1. Do the several permeant ions share a common pathway through the membrane, or are the pathways for different ions different?
2. Is the ionic permeability a diffuse property of broad areas of membrane, or is it localized in rare specializations of the membrane?

To answer these questions in his research for his Ph.D. [34] Hille argued that the fact that Tetrodotoxin, the active element of puffer fish poison, eliminated the Sodium current without effecting the magnitude or the time course of the Potassium current [35], while Tetraethylammonium blocked the Potassium current without influencing the time course or magnitude of the Sodium current indicated two separate molecular mechanisms, or channels as Hille called them, for the Sodium and Potassium currents. Hille's work was also supported by the work of Narahashi and Haas [36], who used the effects of the insecticide DDT (Dichlorodiphenyltrichloroethane) on the membrane potential to arrive at the same conclusions.

The study of channels in the membrane received a significant boost with the development of the patch clamp method by Erwin Neher and Bert Sakmann [37], which enabled the examination of channels in a tiny area of the membrane. For their work, Neher and Sakmann received the Nobel Prize in 1991.

As Hodgkin and Huxley had conjectured, the membrane contains different channels for the transportation of Potassium, sodium, and chloride ions. These channels are transmembrane proteins that selectively allow different ions to pass through. The way the opening and closing, or *gating*, of the channel functions is one way of classifying the channels. The two kinds of channels that are of interest to us are *ligand-gated ion channels* and *voltage-gated ion channels*. There are also non-gated channels which are referred to as *leakage channels*. These channels allow the flow of ions in response to concentration gradients. The channels that Hodgkin and Huxley conjectured are voltage-gated. These channels open and close in response to the potential difference across them. We can further classify these channels based on the types of ions that pass through them. The two voltage-gated ion channels that take part in the action potential are the voltage-gated sodium channel and the voltage-gated potassium channel. The voltage-gated sodium channel can exist in one of three states: closed, open, or inactivated. The gate is closed at the resting potential of -70 mV. It opens at a threshold of about -50 mV and gets inactivated when the voltage across the membrane is significantly more positive than the threshold. The opening and closing actions of the channel occur rapidly. The potassium channels open at approximately the same voltage as the sodium channels, but they are more slowly acting. Ligand-gated ion channels are ion channels that open up when a chemical messenger or *ligand* binds to a receptor on the outward-facing or extracellular part of the ion channel. There is a huge variety of ligands and corresponding receptors—ligands and their receptors usually occur as matched pairs. Binding of the ligand to its receptor leads to a conformational change in the protein's shape, resulting in either the opening or closing of the pore through the membrane.

6.6 Propagation of the Action Potential

Consider the cartoon drawing of the neurons in Fig. 6.5. Our focus is on the neuron on the right. It receives signals from the neurons to the left, integrates these signals, and then possibly transmits the integrated signal to another neuron on the right. Let's begin by examining the delivery of signals from the neurons on the left, which we will refer to as the presynaptic neurons. As we mentioned earlier, the communication across the synapse is chemical in nature. When a signal travels down the axon of the presynaptic neuron, it results in the release of neurotransmitters. These neurotransmitters are packaged into vesicles at the axon terminals. A certain amount of low-level release of these vesicles into the synaptic cleft occurs at all times. However, when the electrical signal in the presynaptic neuron arrives at the axon terminal, it initiates a sequence of actions that results in the fusion of synaptic vesicles with the axon membrane, releasing neurotransmitters into the synaptic cleft. Some of these neurotransmitters bind to receptors on ligand-gated ion channels on the post-synaptic neuron, as shown in Fig. 6.6.

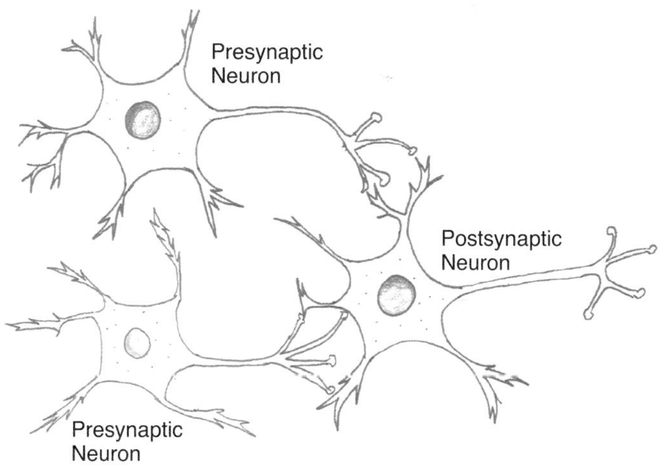

Fig. 6.5 Neurons in a neuronal pathway

Fig. 6.6 Action in the synaptic cleft

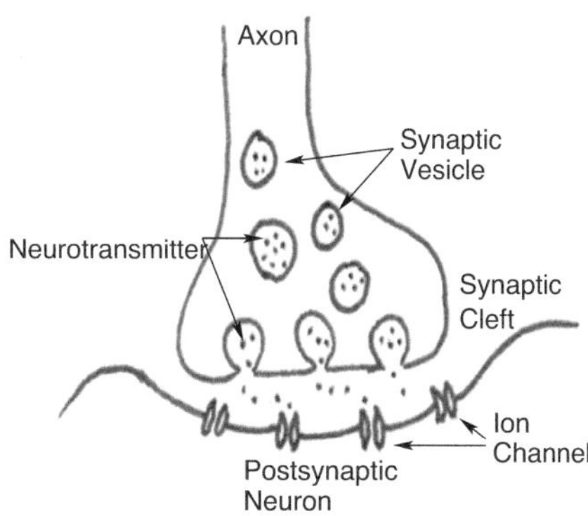

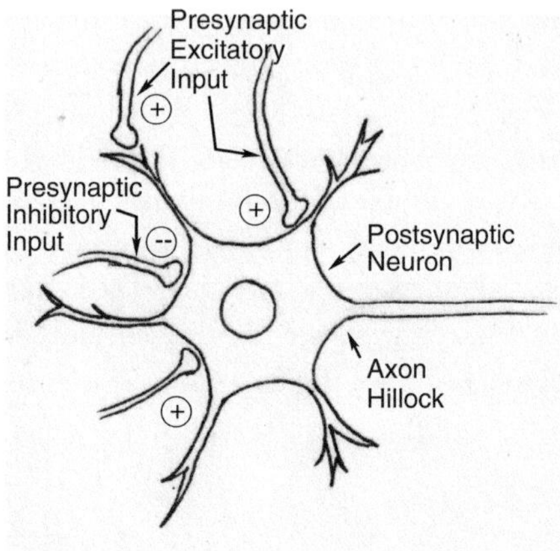

Fig. 6.7 Summation of different inputs to generate the action potential

There are more than 200 kinds of neurotransmitters, including glutamate, gamma-aminobutyric acid (GABA), and acetylcholine. Each of them can either increase or decrease the transmembrane potential by allowing the influx or exit of positively or negatively charged ions. Glutamate, one of the more abundant neurotransmitters, has an affinity for ion channels that are selective for positively charged ions, such as sodium. The release of glutamate into the synaptic cleft will result in the opening of ligand-gated sodium channels. This will result in an influx of sodium ions, which will rapidly increase the transmembrane potential at the location of the sodium channel, making it more positive. This action is called depolarization. When the ligand is released from the receptors, the channel is closed again. The sodium ions diffuse away from the location of the sodium channel carrying the wave of depolarization along the membrane of the dendrite towards the axon hillock. Because this depolarization can result in the generation of an action potential, we say that Glutamate is an excitatory neurotransmitter. GABA, on the other hand, is an inhibitory neurotransmitter. It tends to bind to channels that are selective for negative ions such as Chlorine. The entry of negative chlorine molecules will result in a further depression of the membrane potential, or hyperpolarization, in the vicinity of the channel. These Chlorine ions will also travel towards the axon hillock (Fig. 6.7).

At the axon hillock, the polarization or depolarization effect from multiple ion channels is summed. The axon hillock is the site for a large number of voltage-gated sodium channels. If the summation of both the polarizing and depolarizing signals at the axon hillock results in a membrane potential greater than some threshold value (approximately −50 mV), it will lead to the opening of voltage-gated Sodium channels and a rapid influx of Sodium, signaling the initiation of an action potential. The initiation of the action potential is a binary event. If the membrane potential exceeds the threshold, an action potential occurs, and if the membrane potential is below the threshold, no action potential is generated. The action potential begins with the opening of the fast-acting gated sodium channel. This opening results in a rapid influx of sodium ions, leading to an increase in the membrane potential around the channel.

Fig. 6.8 Action potential

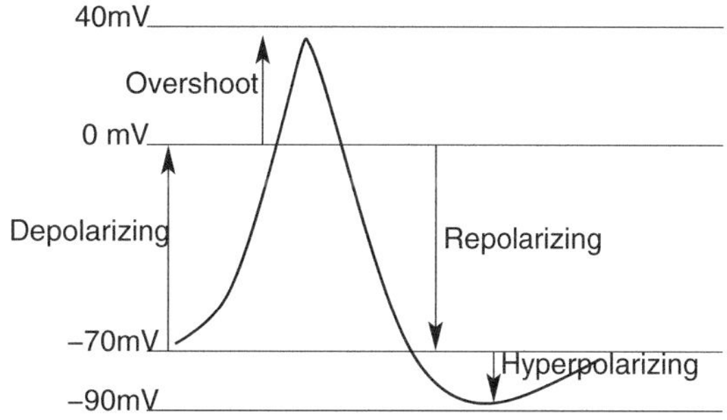

This increase, over time, results in the inactivation of the sodium channel. At about the same time, the slower-acting potassium channel opens, allowing for the outflow of potassium ions that reduces the membrane potential and leads to depolarization. This process continues until the membrane potential drops below the resting potential, at which point the channels enter a refractory period. An example of an action potential is shown in Fig. 6.8.

At this point, the set of channels that took part in this first action potential has entered a refractory period. However, the axon is lined with voltage-gated ion channels. The initial action potential at the axon hillock results in the depolarization of the membrane potential near the neighboring channels, which then initiates their own action potential, progressing through depolarization, hyperpolarization, and finally entering a refractory state. This then triggers the channels further down the axon, and this domino effect looks like the action potential traveling down the axon. Unlike a current in a wire, we do not have electrons flowing down a metallic riverbed. Instead, we have a sequence of action potentials carrying the information from the triggering of the first action potential down the neuron. The movement of action potentials is unidirectional due to the refractory period. After a set of channels takes part in an action potential, they enter a dormant state, which prevents them from being retriggered. This ensures that the propagation of action potentials is always away from the axon hillock.

6.7 Summary

The neuron is the building block on which much of the communication in the human body is built. Despite its fundamental importance, the discovery was relatively recent, as throughout history, thinkers such as Aristotle, Descartes, and Newton viewed neuronal transmission as the flow of animal spirits [38]. As a crucial component of the nervous system, there is much to learn about the neuron, and perhaps even more to discover. In this chapter, we have barely scratched the surface. We looked at the history of our understanding of how neurons work. In particular we looked at the Hodgkin-Huxley model for neuronal membranes and how the channels in the membrane contribute to the generation of action potentials. Our primary goal in this chapter is to demonstrate that the transmission of information across neurons differs significantly from how information is transmitted in other situations with which we are more familiar. Understanding this will help us appreciate the signals we will study in the next two chapters.

The Electrocardiogram 7

7.1 Structure and Overview of the Heart

There are several ways to analyze a heart. To understand its functioning and the connection of this function to the EKG trace, we can use two approaches. We can analyze the heart as if it were composed of two pumps, or we can analyze it as if it were comprised of two semi-isolated electrical systems. Let's look at each method of analysis in turn.

7.1.1 The Heart as a Set of Pumps

The heart comprises four chambers: two atria and two ventricles, with the two atria positioned on top of the ventricles. One of the pumps consists of the right atrium and right ventricle (shown in blue in Fig. 7.1), and the other consists of the left atrium and left ventricle. These pumps are enclosed in a *serous membrane*. A serous membrane consists of two layers: an inner, or *visceral layer*, and an outer, or *parietal layer*. The cells that make up the serous membrane contain glands that secrete a lubricating fluid called the *serous fluid*, which is present between the two layers. This arrangement enables nearly frictionless movement for organs enclosed by the serous membrane. As these organs are packed tightly in our body, this is a necessity. In the serous membrane that encloses the heart, the inner visceral layer is known as the *visceral pericardium*, and the outer parietal layer is referred to as the *parietal pericardium*. The serous fluid between these two layers is called the *pericardial fluid*. Outside the serous layer is a fibrous layer that protects the pumps and is fused to the parietal pericardium. The serous layer and the fibrous layer together are called the *pericardium*.

The atria and the ventricles go through a cycle of contraction and relaxation called *systole* and *diastole*. The period of relaxation, or diastole, allows the chamber to fill with blood, and the contraction, or systole, forces the blood out of the chamber. The deoxygenated blood from the body flows into the right atrium through the *inferior* and *superior vena cava* and

K. Sayood, *Biomedical Signal Processing*, Synthesis Lectures on Biomedical Engineering,
https://doi.org/10.1007/978-3-032-08855-0_7

Fig. 7.1 Human heart

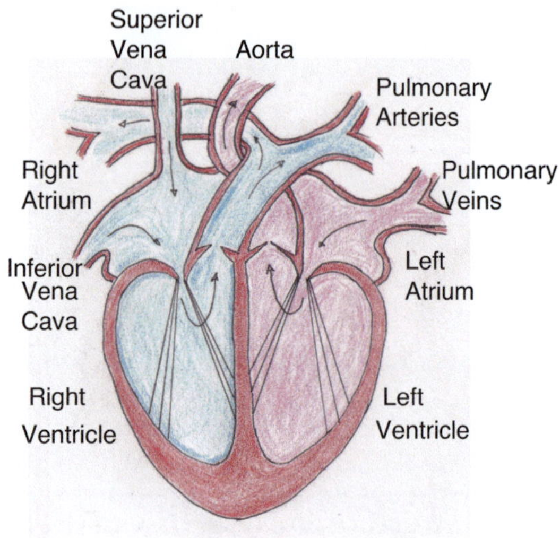

the *coronary sinus*. The inferior vena cava carries deoxygenated blood from the lower part of the body to the right atrium, the superior vena cava carries deoxygenated blood from the upper part of the body to the right atrium, and the coronary sinus collects deoxygenated blood from the cardiac muscles and deposits this blood in the right atrium. About 75% of this blood passively flows from the right atrium to the right ventricle through the tricuspid valve even before the right atrium contracts. The remaining 25% of the blood is pushed into the right ventricle with atrial systole. This blood being pushed in causes the ventricle to expand, which gives it an extra oomph when it contracts. The ventricular systole pushes the blood to the lungs through the pulmonic valve and the pulmonary arteries. After oxygenation in the lungs, the oxygenated blood returns through the pulmonary veins to the left atrium and, from there, into the left ventricle through the mitral valve. Again, the atrial systole is only responsible for about a quarter of the blood flowing into the ventricle. The blood is pushed out of the left ventricle through the aortic valve into the aorta. All this contraction and relaxation happen because of action potentials in the cardiac muscle cells, which leads us to the second way of analyzing the heart.

7.1.2 The Heart as a Set of Electrically Connected Regions

To understand the electrical properties of the heart, we begin with the cells that make up the cardiac muscle. The cardiac muscle cells or cardiac myocytes (or cardiomyocytes) are of a type known as *syncytium*. These cells are connected through *gap junctions*, which directly connect the cytoplasm between cells. In the current context, we can think of these junctions as low-resistance barriers that allow the flow of ions, and hence action potentials, along the cardiac muscle cells connected through gap junctions. While the two atria and the two ventricles are electrically connected in this way, this is not true of the atria and ventricles. Between the two atria and the two ventricles are valves made up of fibrous tissue. As these tissues do not conduct ionic current, this effectively divides the heart into two somewhat separate electrically conducting regions: the atria form one conducting region, and the ventricles form the other. The two conducting regions are insulated from each other by the fibrous tissue.

Instead of the propagation of signals through gap junctions, from muscle cell to muscle cell, signals from the atria go to the ventricles through a bottleneck of conducting fibers. These conducting fibers originate in a group of specialized cells in the endocardium of the inter-atrial septum called the *atrioventricular node*, better known as the AV node. From the AV node, the signal is carried to a bifurcation point with the super cool name—the *bundle of His*, named after its discoverer Wilhelm His Jr.. The bundle of His bifurcates into the left and right bundle branches, which in turn innervate Purkinje fibers distributed to the ventricular muscle.

With this brief description of the heart's major components, let's briefly examine its functioning. The process of systole and diastole is initiated by the cells in the *sinoatrial node* (SAN). The signal from the sinoatrial node spreads through the atria preferentially through electrical conduction tracts called *Bachmann's bundle* and the anterior, middle, and posterior tracts, which converge near the atrioventricular (AV) node. From there, it propagates to the ventricles through the bundle of His and the right and left bundle branches. Let's first examine the action of the sinoatrial node, and then we will consider how the signal progresses through the cardiac muscle cells.

7.2 The Sinoatrial Node

Several cells in the heart have the potential to act as pacemakers. In practice, the role is taken by the sinoatrial node. Unlike most neuronal cells, which enter a refractory period after firing, the cells in the sinoatrial node can self-excite. This periodic excitation enables them to produce the heartbeat rhythm with which we are so familiar. To provide this self-excitation, the neurons in the sinoatrial node are believed to employ several mechanisms. One is an ion channel called the *Hyperpolarization-activated Cyclic Nucleotide-gated* (HCN) channel, and the other is through the sequestration and release of calcium ions. This latter mechanism is sometimes referred to as the calcium clock. Another channel involved in the process is the sodium-calcium exchanger (NCX), a type of ion exchange channel. The sodium-potassium pump maintains the balance between intracellular and extracellular sodium and potassium, thereby polarizing the membrane potential. The sodium-calcium exchanger exchanges one calcium ion (Ca^{2+}) from inside the cell for three sodium (Na^{+}) ions from outside the cell, thus depolarizing the membrane potential.

As their name implies, the HCN channels open during hyperpolarization. Thinking back to the chapter on neuronal communications, recall that during an action potential, once depolarization takes place, the ion channels enter a refractory phase in which they cannot be excited. This prevents the same ion channel from opening again, allowing for the unidirectional propagation of the action potential. At the same time, this also prevents self-excitation. In the cells of the sinoatrial node, the HCN channels remain open during hyperpolarization, rather than becoming inactive, allowing an inflow of both sodium and calcium ions. This ionic current, involving both potassium and calcium, was unexpected at the time of its discovery, so people started calling it the *funny current*, and the name stuck. The HCN channels open at around −60 mV. When the inflow of sodium and calcium ions increases the membrane potential to around −50 mV, calcium channels called the T-type calcium channels open, further depolarizing the membrane. At around −40 mV, the long-lasting, or L-type, calcium channels open, a phenomenon we will explore further later. This entry of calcium triggers another mechanism.

Recall that the sodium-potassium pump maintains the balance between intracellular and extracellular sodium and potassium, thereby polarizing the membrane potential. Another pump that exchanges ions is the sodium-calcium exchanger, which exchanges one calcium ion (Ca^{2+}) for three sodium ions (Na^{+}), thereby depolarizing the membrane potential. Calcium ions are present both outside the cell and inside the cell. Inside the cell, they are contained within compartments known as the *sarcoplasmic reticulum*. The sarcoplasmic reticulum is being constantly filled with calcium by the action of the *Sarcoplasmic Reticulum Calcium ATPase* (SERCA). The increase in calcium concentration in the sarcoplasmic reticulum is detected by calcium receptors, triggering the release of calcium in spurts within the cell. The release of calcium ions from the sarcoplasmic reticulum increases the activity of the sodium-calcium exchanger, thereby enhancing depolarization. It also results in the opening of L-type Calcium channels, which allow calcium to follow its concentration gradient into the cell, leading to further depolarization. The repolarization process begins with the closure of these calcium channels and the subsequent opening of potassium channels. Once the membrane potential drops to approximately −60 mV, the HCN channels reopen, and the cycle repeats.

The cells of the sinoatrial node undergo this periodic activity due to their unique internal structure. However, the length of the period or the rate of the pacemaker can depend on several external signals. Cyclic nucleotides, particularly cAMP, influence the speed of the HCN channel. The rate is also affected by neuronal connections from both the sympathetic and parasympathetic systems. The vagus nerve provides the parasympathetic input. The neurotransmitter released by the vagus nerve, acetylcholine, binds to receptors on the cells of the sinoatrial node, which opens the potassium channels and depresses the funny current. This reduces the slope of the activation and, thus, the rate of the pacemaker cells. The opposite effect is caused by the release of the neurotransmitter norepinephrine by the sympathetic system. Both these neurotransmitters have a wide-ranging impact on the body, of which the cardiac response is a part. Norepinephrine is part of the fight-or-flight response, which includes increased cardiac activity.

From the sinoatrial node, the pacemaker action potential is transmitted in two directions—to the cardiac muscle cells of the atria and, through specialized cells to the atrioventricular (AV) node. At the AV node, there is a delay of about 90 ms. This delay allows for the atrial contraction to complete before the ventricular contraction begins. The action potential travels from the AV node through the bundle of His, the right and left bundle branches, to the Purkinje fibers. If the action potential were to travel only through the conductive fibers, the effect would not be very significant, and it would undoubtedly be challenging to detect from outside the body. Of course, the action potentials are not restricted to these channels. Through the action of gap junctions, they travel through the cardiac muscle cells, also known as cardiac myocytes. The spread of action potentials causes ventricular contraction, which is what is detected externally.

7.3 Myocyte Action Potential

The action potential in cardiac muscle cells differs from that of the sinoatrial node. It involves fast-acting sodium channels, slow calcium channels, and the ever-present potassium channels. As shown in Fig. 7.2, the action potential is divided into five phases: phases 0, 1, 2, 3, and 4. The resting potential of the myocyte cells is approximately −90 mV, slightly higher than the equilibrium potential of potassium at −92 mV. When the neighboring cell undergoes depolarization, some positive ions cross the gap junctions, initiating phase 0 of the action potential. In this phase, the potential initially rises due to the leakage of positive ions through the gap junction. Once the potential reaches about −70 mV, the fast-acting sodium

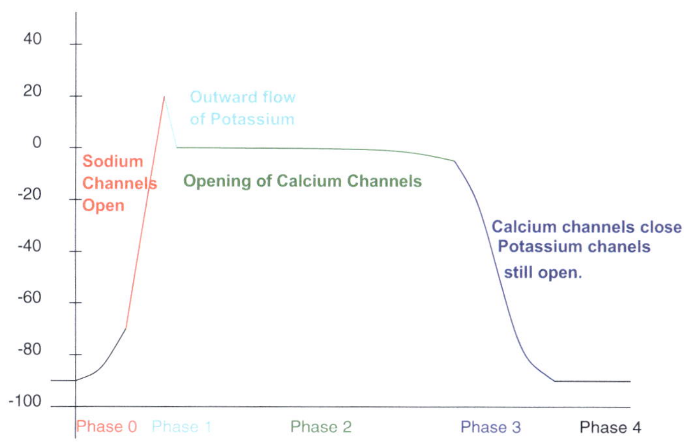

Fig. 7.2 Myocyte action potential

channels open rapidly, depolarizing the membrane. At the same time, the much slower-acting calcium channels also begin opening. The fast-acting sodium channels are open only for a thousandth of a second. When they close, the effect of the potassium channel begins the process of depolarization. This is the first phase of the action potential. By this time, the slow-acting calcium channels are open, counteracting the effect of the potassium leakage, and for a while, the membrane potential remains constant at around zero mV. This is the second phase of the action potential. The slow-acting calcium channels finally close, and the effect of the potassium channels leads to a polarization of the myocyte cell membrane. This is phase 3. During this phase, the excess calcium is removed by the sodium-calcium exchanger and SERCA. Finally, the cell membrane returns to its resting potential, or phase 4, at approximately −90 mV.

Note that the fast-acting sodium channels only open for a short period, after which they enter a refractory period until stage 4 of the action potential. It appears that these channels require the really low membrane potential of stage 4 to reset. If the membrane potential does not reach these low levels, as is the case in the pacemaker cells, the fast-acting sodium channel stays closed.

7.3.1 Contraction of the Myocytes

We won't try to go through the contractile process in any detail. Here is a very simple story, illustrated by a cartoon drawing of a component of the muscle cell known as a *sarcomere* (Fig. 7.3). Let's introduce the set of proteins that will carry out the process of contraction. The protein *myosin* exists as thick filaments with globular heads spaced along the filament. These globular heads contain myosin ATPase, which is used to release the energy required for contraction from ATP molecules. *Actin* molecules are thin filaments that interdigitate with the thick myosin filaments and are bound to the borders of the sarcomere called the *z line*. The myosin is connected to the z line with a springy filament called *Titin*.

The contraction occurs when the myosin globular heads grab the actin filament and pull on it. However, before it can grab the actin filament, the myosin globular heads must first find sites on the actin to which they can hold on. These sites are shielded by a protein called tropomyosin, which in turn is controlled by a protein called troponin, which is attached to the actin fibers. During depolarization, calcium enters the cell and triggers a much bigger release of calcium from the sarcoplasmic reticulum. This phenomenon is known as calcium-

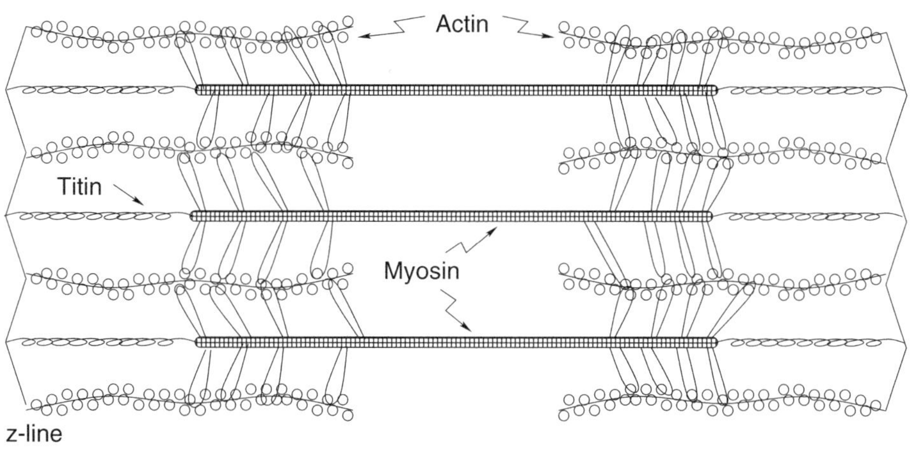

Fig. 7.3 Components of a sarcomere

induced calcium release (CICR). The calcium that floods the cell binds to troponin, which in turn moves tropomyosin away from its shielding position. This allows the myosin globular heads to grab on to the actin filaments and pull on them, thus contracting the cell.

7.4 Review

So, after all of this, here is what we have said about how the heart functions. The sinoatrial node is the heart's pacemaker. It continually generates action potentials, which initially spread to the cardiac myocytes of the atria. The depolarization of these muscle cells results in a contraction of the atria. The action potentials propagate to the AV node, where, after a 90 ms delay, the action potential spreads to the ventricles through the bundle branches and the Purkinje fibers to the ventricular muscle cells. Just as in the case of the atria, the action potentials in the ventricular cardiac muscles cause a contraction of the ventricles.

We have attempted to condense a substantial amount of material into a tiny amount of space. If you want more complete (and coherent) information, a couple of very nice sources are the books *Textbook of Medical Physiology* [39] and *Pathophysiology of Heart Disease* [40].

7.5 The EKG Trace

The cardiac cycle is continuous and cyclical, but the standard convention is that the depolarization of the atria marks the beginning of each cycle. Following this convention, the cycle of heartbeats begins with the depolarization of the atria. This is followed by the depolarization of the ventricles, culminating in the repolarization of both the atria and ventricles. The fact that the functioning of the heart involved electrical potentials and currents was well known in the nineteenth century. Measuring these electrical variations without exposing the heart was another matter. The first person to do so was Auguste Desire Waller. According to him, he had just finished observing the workings of the heart of a decapitated cat when it occurred to him that it might be possible to observe the electrical activity without having to decapitate the patient. He dipped his right hand and left foot into basins of saltwater and connected the basins to a Lippmann electrometer. The Lippmann electrometer detects changes in potential through the movement of mercury in a tube. Waller conducted various experiments on himself, his assistants, and his dog, Jimmy—the latter led to complaints in the British House of Commons [41] (the decapitated cat aroused no such furor). While Waller's approach provided a way of observing gross changes, the electrometer was too slow to reflect all the changes taking place on the surface of the heart. One of the people to whom Waller demonstrated his discovery was a Dutch scientist named Willem Einthoven. Einthoven took Waller's idea and made it practical. Instead of using a Lippmann electrometer, he used a string galvanometer, which was much more sensitive to the changes in electric potential. He developed the methods for attaching leads and much of the theory that is still in use today. Einthoven attached electrodes to a person's arms and left leg and measured the voltages between them.

7.5.1 12 Lead EKG

In the original EKG of Waller and Einthoven, the electrodes were conducting buckets filled with saltwater. The modern electrodes are somewhat smaller. In the standard 12-lead ECG system, we use ten electrodes. Three of the electrode locations are identical to those used by Einthoven. The traces, or leads, they generate are called *limb leads*. How the limb leads are generated is shown in Fig. 7.4. Three leads, Lead I, Lead II, and Lead III, are the potential

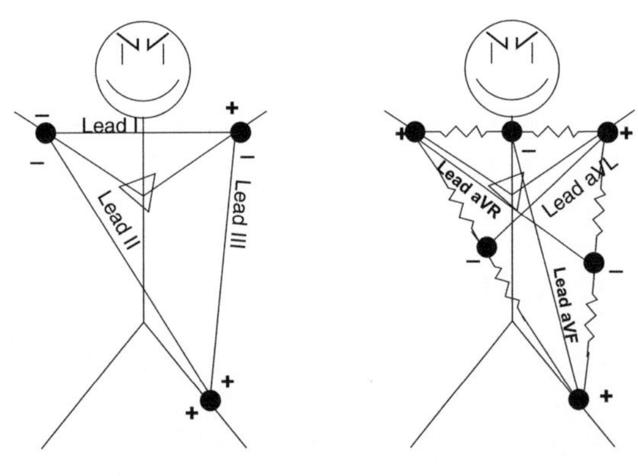

Fig. 7.4 ECG Leads

differences between the three Einthoven electrodes. The other three, aVL, aVF, and AVR, also known as augmented leads, represent the potential differences between a limb and the midpoint between the other limbs. Thus, three electrodes generate six limb leads. The other six leads are generated from electrodes placed on the chest and are referred to as chest leads or precordial leads. They are labeled V1 through V6. Each of these connections offers a different perspective on the same process.

7.5.2 Polarization Sequence and the EKG

Before examining the EKG signal, let's first consider what happens in a simplified membrane during depolarization. We do this by measuring the voltage across the outside of the membrane. Remember that when the membrane is polarized, the inside of the membrane is at a lower potential than the outside of the membrane. During depolarization, this situation is reversed, and the inside of the membrane becomes at a higher potential than the outside of the membrane. If we measure the voltage on the outside of the membrane, we observe a change in potential across the membrane, with the depolarized portion becoming more negative than the polarized locations. We can see this from a simplified one-dimensional drawing of the depolarization process in Fig. 7.5. Notice that the deflection of the voltage signal will be positive or negative depending on the polarity of the voltmeter as well as the direction of depolarization.

As we mentioned earlier, there are several different signals we can get from the heart. We will examine what is commonly referred to as Lead II, in which the voltmeter is connected between the right arm and left leg, with the negative end of the voltmeter on the side of the right arm. When the atria depolarize, we get a deflection similar to the one shown in Fig. 7.5. This is called the *P wave*. The P wave lasts anywhere between 80 and 120 ms. After the P wave, there is a period of 40–100 ms during which the electric potential remains constant. This isoelectric region occurs because the atria have contracted, while the delay in the AV node means that depolarization of the ventricles has not yet occurred. The signal corresponding to the repolarization of the atria gets masked by the much larger signal corresponding to the depolarization of the ventricles. This signal is referred to as the *QRS complex*. The QRS complex consists of a sequence of three deflections. The first negative deflection is called the *Q wave*, the first positive deflection is called the *R wave*, and the second negative deflection is called the *S wave*. Sometimes, this is followed by another positive deflection, known as the *R' wave*. The entire QRS complex takes about 120 ms and occurs about 120 to 200 ms after the P wave. The QRS complex is followed by another isoelectric region, which is followed by the signal corresponding to the repolarization of the ventricles. This constant potential region is thought to correspond to the flat phase 2 region of the cardiac myocyte action potential. This part of the EKG signal is called the T wave, while the isoelectric region between the QRS complex and the T wave is called the ST segment. The various parts of this depolarization-polarization cycle are depicted in Fig. 7.6.

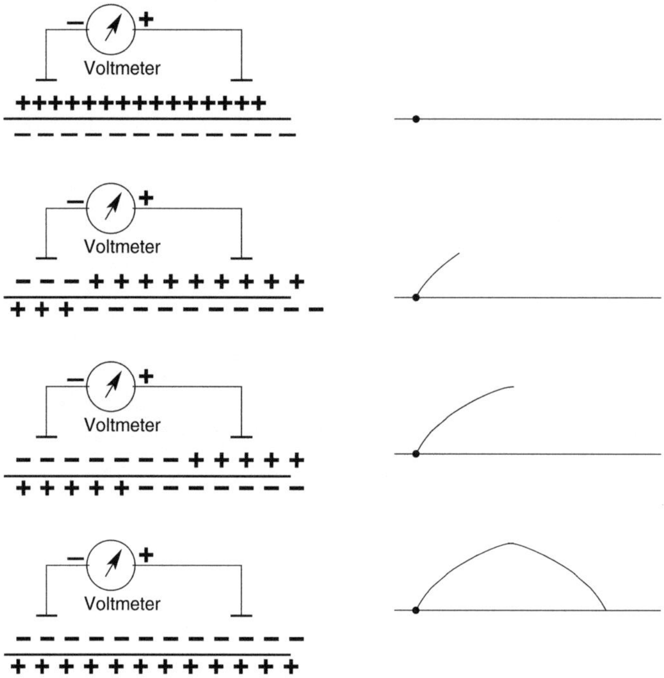

Fig. 7.5 Progression of depolarization

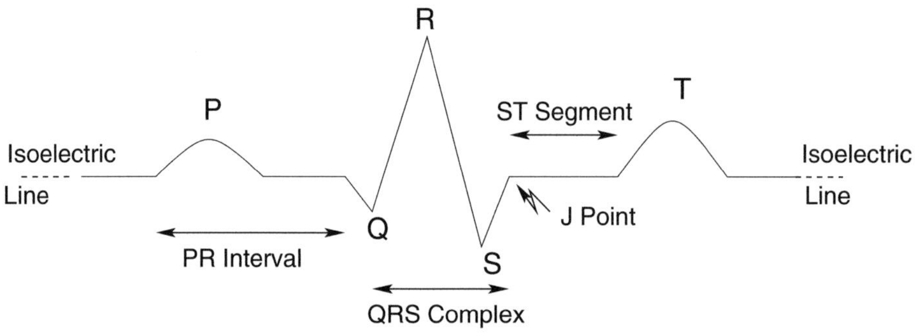

Fig. 7.6 One cycle of an EKG

7.6 Analysis of EKG Signal

By examining the EKG trace, we can gain valuable insights into potential heart issues. The different parts of the EKG trace—the P wave, the QRS complex, the ST segment, the PR segment, and the QT segment—all correspond to distinct events during a cardiac cycle. As such, physiological abnormalities in these events are reflected in variations of these various components. Here is a partial list of the variations in the components of an EKG and the connection of these variations with physiological pathologies. We can see how signal processing techniques can be used to detect variations in the EKG trace, helping clinicians diagnose pathologies.

P-wave The P-wave is a reflection of the depolarization of the atria. It normally lasts between 80 and 120 ms. Deviations from its normal presentation can be indicative of various malfunctions.

- *Wandering Atrial Pacemaker*: The pace of the heartbeat is usually set by the Sinoatrial node (often referred to as the pacemaker of the heart). A normal cardiac rhythm is often referred to as the *sinus rhythm*. The action of the Sinoatrial node can malfunction due to either the node's malfunction itself or due to blocks on the path from the Sinoatrial node to the Atrioventricular node, which also contains pacemaker cells, or by organized activity outside of both that overrides the pacemaking activity of the SA node. When this activity originates from bundles of cells in the atria, it's classified as a wandering atrial pacemaker. On the EKG, this is diagnosed when the P-wave preceding each beat assumes three or more distinct shapes.
- *Atrial fibrillation* In this, the QRS shape is still normal, but in place of the P wave and the flat region between the P wave and the QRS complex, we get rapid, chaotic movement.
- *Junctional rhythm* Alternatively, when the SA node is overridden by activity from the atria that are not organized (i.e., random), it results in an EKG with irregular intervals between QRS complexes but no discernible P-wave (since there is no synchronized contraction of the atria).
- *Atrial flutter* Usually, the depolarization wave spreads quickly enough that it is unidirectional, moving away from the sinoatrial node. However, if there are locations outside the sinoatrial node that are performing pacemaker activity, or if damaged tissue is slowing the spread of the action potential, a circular depolarization path is possible. This results in a 'sawtooth' pattern as the circular depolarization progresses around, between QRS complexes that make it through the AV node. This will lead to atrial tachycardia. In this, the QRS wave is normal, but the distance between the P wave and the QRS complex is shortened considerably. When the distance disappears, this is called *atrial flutter*.
- *Right atrial hypertrophy* This usually happens when the right atrium has to work abnormally hard, perhaps because of some blockage of the aortic valve. This results in an enlarged atrium, which is reflected in an unusually large P wave.
- *Left atrial hypertrophy* The depolarization of the left atrium is reflected later in the P-wave. An increased size of the left atrium, perhaps due to dysfunction of the mitral valve, is reflected in the EKG as a double-humped P wave.

QRS complex Issues with ventricular rhythm are usually reflected in the QRS complex.

- *Premature ventricular contraction* In this situation the QRS complex has an abnormal shape, so you might not detect the R wave. This will result in an RR interval that is different from the normal RR interval.
- *Right bundle branch block* This happens when the signal flow in the right bundle branch is blocked for some reason. This means that the depolarization signal must originate from the left bundle through the myocytes in the left ventricle, which then travel to the right ventricle. This, in turn, means that the first part of the QRS complex is normal, but there is a sharp downward deflection of the S wave. This is most apparent in Lead I and the chest leads.
- *Left bundle branch block* This happens when the signal flow in the left bundle branch is blocked for some reason. In the EKG, this results in longer QRS duration and broad, tall, and sometimes notched R wave. The change in shape in the QRS complex results in a deep S-wave in lead V1 and a comparatively tall R-wave in lead V6.
- *Idioventricular rhythm and ventricular tachycardia* Similar to an atrial pacemaker, this is a pacemaker action that originates outside of the sinoatrial node. This typically results in a wide QRS complex, as the ventricular contraction originates outside the usual conduction system. If P-waves are seen, they are often not associated with the QRS complexes, given the overriding ventricular activity, but can sometimes be seen inverted as the depolarization propagates backward from the ventricles to the atria.
- *Ventricular fibrillation* Similar to atrial fibrillation, this results from disordered and unsynchronized activity in the ventricles that prevents ordered depolarization. This results in chaotic, irregular waves on the EKG signal with no clearly delineated P, QRS, or T waves. Unlike atrial fibrillation, this can be a fatal condition absent immediate intervention and is always associated with cardiac arrest in need of prompt resuscitation

ST segment The ST segment is the flat interval between the S wave and the T wave. This corresponds to phase 2 of ventricular polarization. This is the constant potential region of the myocyte action potential.

- *ST depression* The term ST segment depression is something of a misnomer. In this condition, the ST segment remains unchanged, while the rest of the waveform becomes elevated. So, relatively, the ST segment looks depressed. The reason for this change in the waveform is a partial blockage of the coronary arteries. This results in an interruption of oxygen supply to the cardiac muscle cells, internal to the endocardium, leading to cell death and the release of potassium into the extracellular space. This, in turn, causes the potential differences during the EKG cycle to rise relative to the isoelectric line. The ST segment is not affected, resulting in an apparent ST depression.
- *ST elevation* Complete, as opposed to partial, blockage of coronary arteries leads to ST elevation. As in the case of ST depression, what happens here is that the EKG waveform, except for the ST segment, is depressed relative to the isoelectric line. This appears to be an elevation of the ST segment.

A more detailed description of what happens during ST elevation and depression is presented in [42, 43].

T wave The T wave is due to the phase 3 rapid polarization of the ventricles. They are normally upright, as shown in Fig. 7.6, in leads I, II, aVL, aVF, and V2-6, and generally inverted in aVR. In leads III and V1, the T-wave can be upright or inverted. Because the T-wave is a reflection of the repolarization of the ventricles, it reacts strongly to changes in conduction in the cardiac muscles. In particular, abnormal increases and decreases of potassium are reflected in changes in the shape of the T-wave

- *Hyperkalemia* This is a life-threatening medical condition caused by an elevated concentration of potassium in the serum. The elevated potassium levels change the conduction characteristics in the cardiac muscles during repolarization, resulting in a *peaked T-wave*. If potassium levels continue to increase, the QRS complex widens and the T-wave disappears in a QRS complex-T-wave fusion.
- *Hypokalemia* This is the opposite condition resulting from an abnormally low level of potassium in the serum. As might be expected, this results in a flattening or inversion of the T-wave. This is accompanied by increased P-wave amplitude, increased PR interval, and ST depression.

There are also T-wave changes in other conditions, but they are generally secondary to the changes in other parts of the EKG. For example, in myocardial ischemia, there is a flattening or inversion of the T-wave in conjunction with an ST-segment depression.

PR interval: The interval between the P-wave and the QRS complex is usually around 120–200 ms. Any blockage of the communication path between the atria—reflected in the P-wave—and the ventricles—reflected in the QRS complex-results in an increase of time for the signal to travel from the Sinoatrial node to the AV node and beyond, which, in turn, results in an increase in the PR interval.

- *First-degree AV block* This is not really a block but an abnormal increase in the PR interval to over 200 ms due to a dysfunction in the AV node.
- *Second-degree AV block* There are two types of second-degree AV blocks, Mobitz Type I and Mobitz Type II. In Mobitz type I, there is a dysfunction in the AV node that prevents it from repolarizing properly between impulses arriving from the sinoatrial node. This is reflected in the EKG by a prolonged PR interval and a periodic skipping of the QRS complex. The Mobitz type II block usually results from blockage in the neural pathways downstream from the AV node in either the bundle of His. As with the type I block, it is reflected in the EKG by the sporadic disappearance of the QRS complex. The difference between Type I and Type II EKG presentations lies in the length of the PR interval. In the case of Type I, the PR interval changes, whereas in the case of Type II, the PR interval (or the PP interval, if the QRS complex is not available) remains constant.

- *Third degree AV block* In this block, there is no communication between the atria and the ventricles. This is a total block of conduction at either the AV node, the bundle of His, or the bundle branches. It is reflected in the ECG by a normal PP interval, but total desynchronization between the P and QRS complexes.

QT interval The QT interval is the time between ventricular depolarization and repolarization. Its length is approximately 270 ms, which corresponds to the duration of the myocyte action potential. Therefore, it is affected by conditions that change the length of this interval.

- *Hypocalcemia* Abnormally low calcium levels entering the cardiac myocyte result in a prolongation of phase 2 of the myocyte action potential. This results in an extension of the ST interval, and therefore the QT interval.
- *Hypercalcimia* The opposite situation occurs with an abnormal increase in calcium levels, which results in a shortening of the ST segment and hence the QT interval.

Clearly, the EKG signal contains a great deal of information. We have gone through many possible pathologies in a somewhat rapid-fire manner. If you are interested in a particular abnormal rhythm, Chap. 19 of the book by Mamivuo and Plonsey [44] contains informative illustrations of the EKG traces for many of these conditions. With the availability of consumer devices capable of recording single leads, signal processing techniques can be applied to provide timely and accurate information to screen and alert people to possible problems [45, 46]. Given the usefulness of the EKG signal, we should ensure that it is processed in a manner that preserves the information it carries.

7.7 Processing of the EKG Signal

The digitization of the EKG signal has opened up a variety of possibilities for both improving the signal through preprocessing and for clinical decision support. The EKG signal can be contaminated in several ways. This can include power line interference, which appears as a 60 Hz component in the EKG signal; baseline wander due to motion and breathing; artifacts resulting from muscle contraction; and other types of electronic interference.

Power line interference Because of the possibility of strong electromagnetic interference from power lines, most EKG systems incorporate amplifiers with strong common-mode rejection. Despite this, contamination can still occur due to cables or poor electrode connections. These show up as signals at the power line frequency and at harmonics of that frequency. This rapidly varying signal can sometimes be misinterpreted as atrial fibrillation; therefore, it is essential that it be removed. The easiest way to remove this would be to use notch filters at the frequency of the power line and its harmonics. However, these can introduce distortions of their own, especially as the interference is nonstationary. To combat this, we can use adaptive filters. Another popular approach is to estimate the noise and then remove it from the signal.

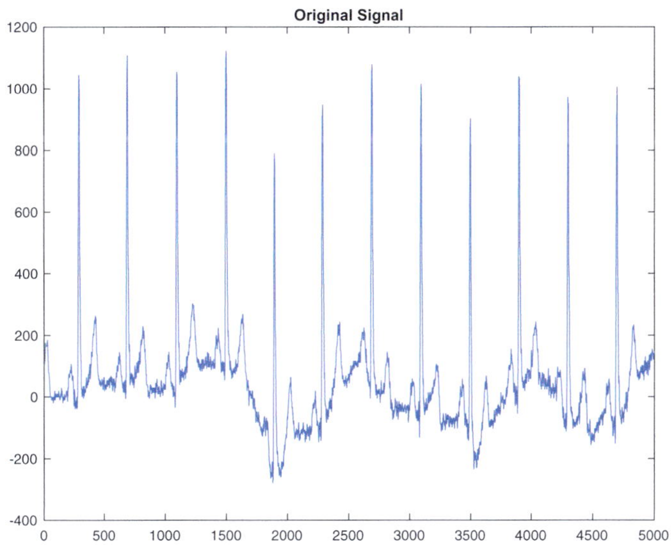

Fig. 7.7 Example of Lead II EKG trace with baseline wander

Baseline wander Baseline wander is exactly what the name suggests. Due to motion or breathing, the baseline of the EKG is unstable and tends to fluctuate, or wander around. An example of baseline wander is shown in Fig. 7.7.

The baseline wander is especially problematic for clinical decision support systems. Most computational approaches require an estimate of the various distances between critical points in the EKG. In particular, the R-R interval—the distance between consecutive R peaks is of particular use. The baseline wander makes it challenging to identify the R peaks in the EKG. Noting that the wander is a low-frequency component, we can attempt to filter the EKG trace with a high-pass filter. The simplest high-pass filter would be a one-pole, one-zero system with the zero at $z = 1$, and the pole close to the unit circle on the real number line.

$$H(z) = \frac{z - 1}{z - 0.99}$$

From this transfer function, we can obtain the difference equation

$$y[n] = x[n] - x[n-1] + 0.99y[n-1]$$

If we filter the EKG trace shown in Fig. 7.7 with this filter, we get the result shown in Fig. 7.8.

This is slightly better in terms of the baseline wander, but the baseline wander is still there. What if we try using a two-pole two-zero filter? We will leave the double zero at $z = 1$ but put the poles, p and p^*, above and below the axis at an angle of 0.05 radians. This results in an input-output relationship

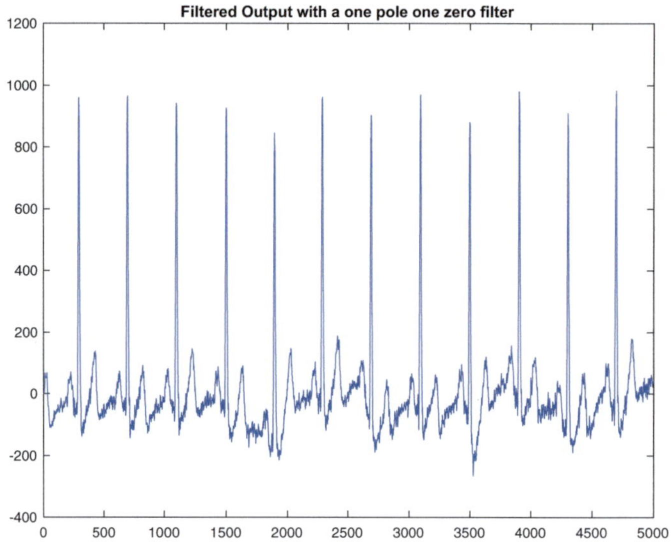

Fig. 7.8 Result of filtering the EKG trace using a one-pole one-zero filter

$$y[n] = x[n] - 2x[n-1] + x[n-2] + \alpha y[n-1] - \beta y[n-2]$$

where

$$\alpha = 2|p|\cos(0.05)$$
$$\beta = |p|^2$$

The result of this filtering is shown in Fig. 7.9. The baseline wander problem is gone! However, we have introduced significant distortion into the signal. This may or may not be a problem. If we are only interested in finding the R peaks, the distortion present in the rest of the signal is irrelevant. However, if that is not our goal, perhaps we should look at other solutions.

One possible solution is to take the FFT of the EKG trace, zero out the low-order coefficient, and reconstruct the signal. If we take the FFT of the trace and set the first five coefficients to zero, the resultant waveform is shown in Fig. 7.10.

The baseline wander is still present, but it appears to be improved. If we now set the first 50 coefficients to zero, we obtain a slightly better result, as shown in Fig. 7.11.

In doing so, have we introduced any distortions? To see that, let's take a closer look at one cycle. As we can see from Fig. 7.12, there is some distortion introduced into the waveform. In particular, we observe a deepening of the S-wave and ST elevation, both of which could impact diagnostic performance. However, again, it is essential to look at the endpoint application. For most applications, this is a very reasonable solution.

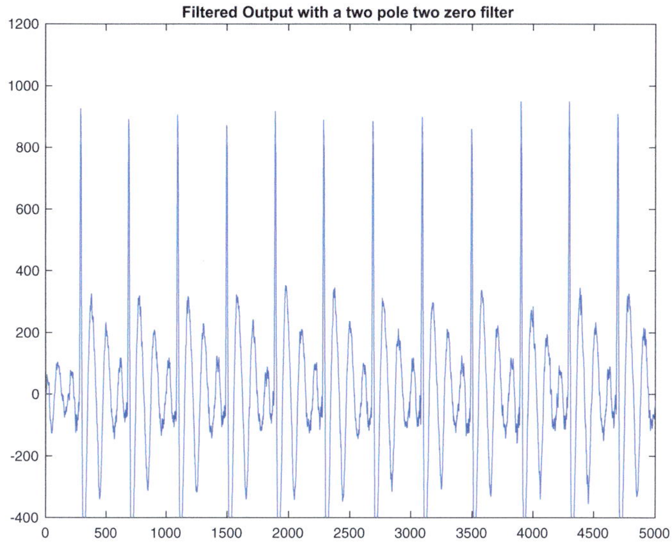

Fig. 7.9 Result of filtering the EKG trace using a two-pole two-zero filter

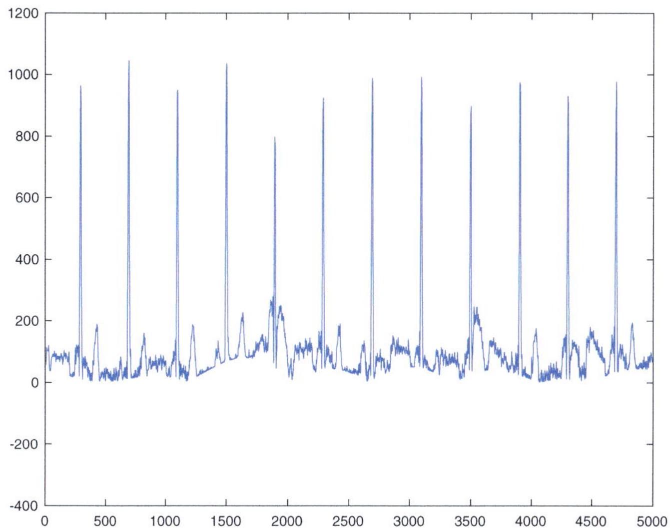

Fig. 7.10 Result of zeroing out the first five coefficients of the FFT of the EKG trace

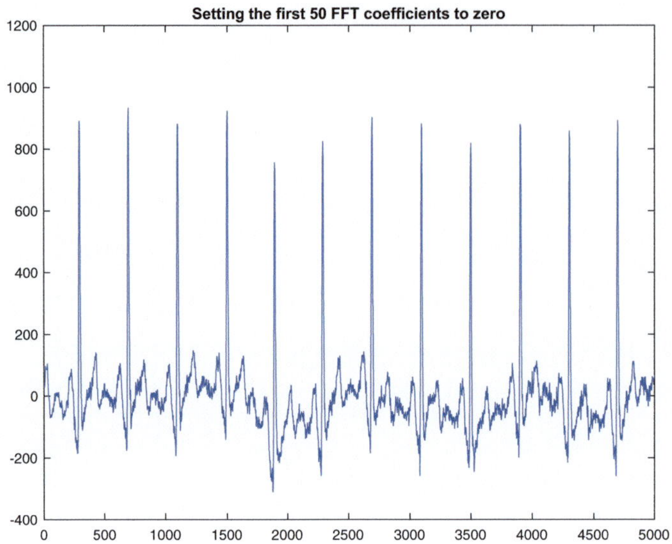

Fig. 7.11 Result of zeroing out the first fifty coefficients of the FFT of the EKG trace

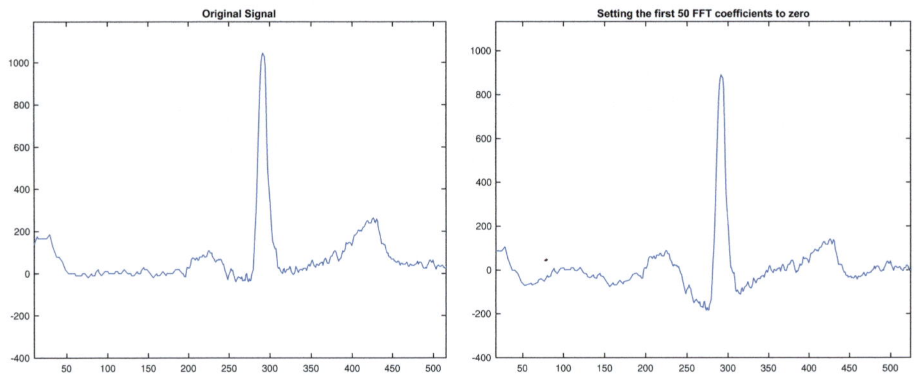

Fig. 7.12 Comparing the original and FFT filtered ECG traces

7.7.1 QRS Extraction

The QRS complex is a dominant feature of the EKG. As such, identifying its location becomes an important task. Several algorithms have been proposed for this purpose. We describe one of the most popular ones- the algorithm proposed by Pan and Tompkins in 1985. The Pan-Tompkin approach can be summarized as shown in Fig. 7.13.

Let's look at each of these blocks in turn.

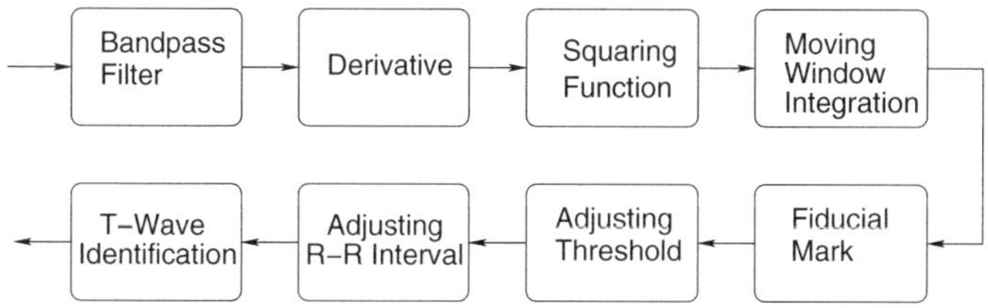

Fig. 7.13 A block diagram of the Pan-Tompkin algorithm for QRS extraction

7.7.1.1 Bandpass Filter

The bandpass filter in the Pan-Tompkins approach is designed to remove all the different sources of interference we have mentioned—power line interference, baseline wander, and muscle noise. To achieve this, they utilize a bandpass filter, which is essentially the combination of a low-pass filter and a high-pass filter. The goal is to isolate the 5–15 Hz frequency range, which contains information about the QRS waveform. In their paper, they present a specific low-pass and high-pass filter. However, this is specific to their sampling rate. The nice thing about their filters is that they all have integer coefficients, making them easy to implement.

7.7.1.2 Derivative

Once the signal has been cleaned, a five-point derivative is used to obtain the slope information for the QRS waveform. The input-output relationship for the derivative used in the paper is given by

$$y[n] = \frac{1}{8T}\left[-x[n-2] - 2x[n-1] + 2x[n+1] + x[n+2]\right]$$

7.7.1.3 Squaring Function

The derivative function output will mark the rise and fall of the QRS waveform with larger values than the surrounding areas, as hopefully most of the noise will be removed due to filtering, and the remaining ripples will have small derivatives. This includes the slopes for the P-wave and the T-wave. The slope will be positive on the rising edge of the QRS waveform and negative on the falling edge. We will use a threshold to determine the extent of the QRS waveform. To simplify the task, we can square the output of the derivative operation.

7.7.1.4 Moving Window Integration

The next step is to integrate the squared derivatives. This ultimately enhances the principal peaks and smooths out the signal. The window for integration is approximately one QRS interval, which Pan and Tompkins take to be 150 ms. The number of samples used in the integrator will depend on the sampling frequency.

$$y[n] = \frac{1}{N}\sum_{k=0}^{N-1} x[n-k]$$

where N depends on the sampling frequency. The selection of N should be done with care, as a large value of N could merge the QRS complex and the T wave.

7.7.1.5 Fiducial Mark

The reference or fiducial mark depends on the application. The QRS complex itself corresponds to the rising portion of the integrated waveform.

7.7.1.6 Adjusting the Thresholds

The thresholds used for noise rejection and peak detection in the Pan-Tomkins approach are adaptive. A peak is detected when the waveform goes from rising to falling in a predetermined interval. This peak could be a signal peak or a noise peak. Determining which one it, is a function of the adaptive threshold algorithm. There are two thresholds: one to detect a signal peak and one to detect a noise peak.

Let S_{peak-i} be the estimate of the signal peak, and N_{peak-i} be the estimate of the noise peak in the integrated signal and S_{peak-f} and N_{peak-f} be the estimates of the signal and noise peaks in the filtered signal. Let Ti_1 and Ti_2 be two thresholds that we will use for the integrated signal. When we detect a peak, we compare it against Ti_1 to determine if it is a signal peak. If no signal peak is detected within a specified time interval, it is assumed that a QRS complex was missed, and we search backward for a signal peak using Ti_2. If a peak is detected and it falls below Ti_1, we initially label it as a noise peak; if it exceeds the threshold, we initially label it as a signal peak. We then update the values of S_{peak-i}, or N_{peak-i}, depending on whether the peak has been classified as a signal peak or a noise peak. If the peak has a value Pi and it is classified as a signal peak, we update the estimate of the signal peak

$$S_{peak-i} = 0.125Pi + 0.875S_{peak-i}$$

If the peak is classified as a noise peak, the estimate of the noise peak is updated as

$$N_{peak-i} = 0.125Pi + 0.875N_{peak-i}$$

In either case the thresholds Ti_1 and Ti_2 are updated as

$$Ti_1 = N_{peak-i} + 0.25\left(S_{peak-i} - N_{peak-i}\right)$$

$$Ti_2 = 0.5Ti_1$$

Once a peak has been identified as a signal peak in the integrated signal, we return to the filtered EKG signal, locate the peak there, and compare the putative peak against a threshold Tf_1. If the peak here is also identified as a signal peak, then this is accepted as a valid QRS complex. The estimates of the signal and noise levels in the filtered signal are updated in the same way as they were for the integrated signal. If the peak is a signal peak with value Pf the estimate of the signal level is updated:

$$S_{peak-f} = 0.125Pf + 0.875S_{peak-f}$$

If the peak is identified as a noise peak, the estimate of the noise level is updated as

$$N_{peak-f} = 0.125Pf + 0.875N_{peak-f}$$

The thresholds are updated as

$$Tf_1 = N_{peak-f} + 0.25\left(S_{peak-f} - N_{peak-f}\right)$$
$$Tf_2 = 0.5Tf_1.$$

7.7.1.7 Adjusting the RR Intervals

Once we have identified the peaks, we have identified the R locations. From these identified locations, we can obtain the R-R interval. Based on the length of the R-R interval, we can determine whether we have correctly identified the QRS complexes. Did we miss one? Did we identify a peak as an R peak when it wasn't actually an R peak? We can determine this by checking if the R-R interval falls within acceptable limits. To do this, the Pan-Tompkin algorithm maintains two R-R interval averages. The first average, let's call it $\bar{RR}_1$, is simply the average of the last eight R-R intervals. From among the R-R intervals, we select those intervals about which we are certain that they have been correctly identified. The second R-R average, $\bar{RR}_2$, is the average of the last eight of these certain R-R averages. An R-R average is defined to be certain if it falls within 92% and 116% of $\bar{RR}_2$. If a QRS complex is not identified within 166% of $\bar{RR}_2$, then the algorithm decides that a QRS complex has been missed, and the most prominent peak between $0.92\bar{RR}_2$ and $1.16\bar{RR}_2$ of the last R peak is considered to be an R peak. If all the previous eight peaks satisfy the condition of falling between 92% and 116% of $\bar{RR}_2$, then we can assume that the heart is beating normally. The rhythm is a normal sinus rhythm.

7.7.1.8 T-Wave Identification

Another possible mistake in this process is when a T-wave is identified as a QRS complex. We can detect this if an R-R interval is smaller than 360 ms. If this happens, we check the slope of the process. If it is less than half the previously identified QRS complex, then we relabel it as a T-wave.

7.8 MATLAB Implementation of QRS Detection Using Wavelets

There is a MATLAB implementation of the Pan-Tomkins algorithm by Hooman Sedghamiz [47] as a function *pan_tomkin(ecg, fr, gr)*, where *ecg* is a one-dimensional array containing the ECG signal, fs is the sampling frequency, and *gr* is a flag which is set to 1 when you want to plot the output of the various stages of the algorithm. The function is available on the MathWorks File Exchange. A sample output of the program is shown in Figs. 7.14 and 7.15.

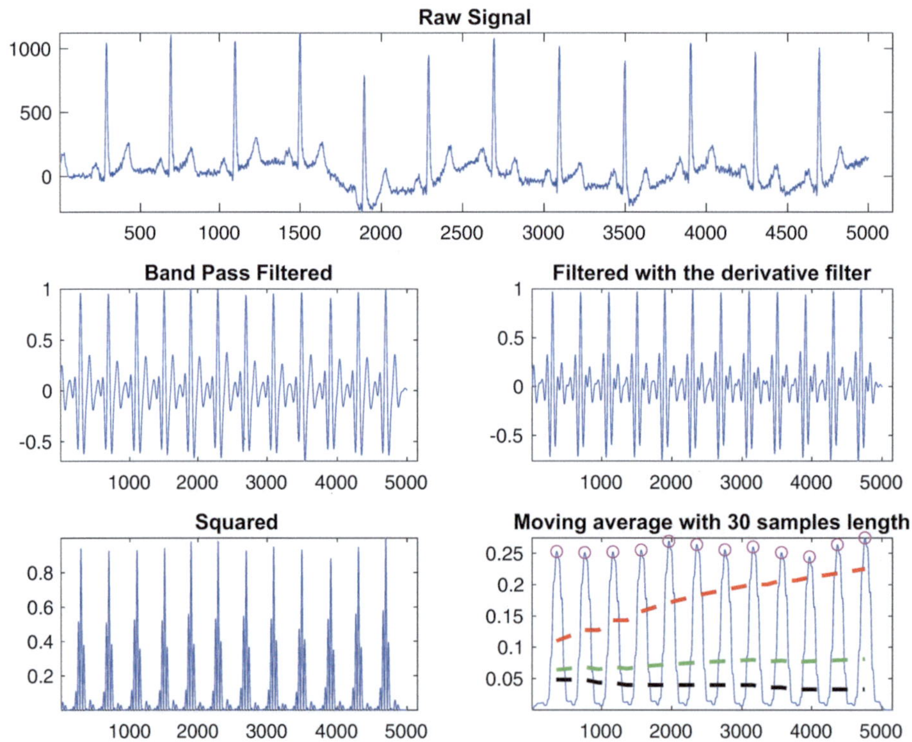

Fig. 7.14 Output from the stages of the Pan Tompkins algorithm

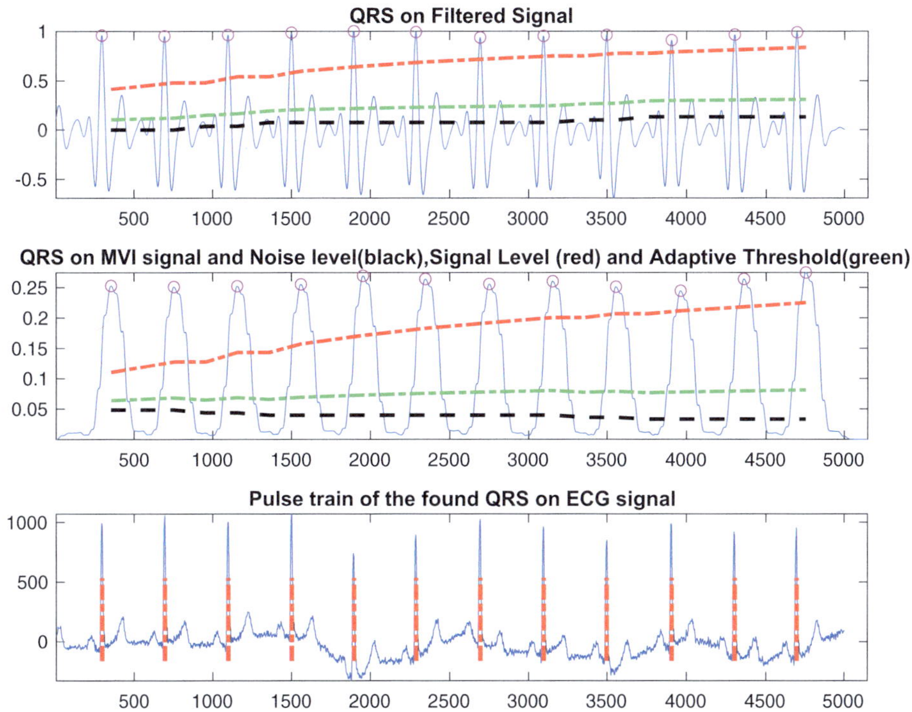

Fig. 7.15 Final outputs of the Pan Tompkins algorithm

7.9 Summary

Our goal in this chapter is to introduce the EKG signal and discuss some standard signal processing applications. To understand the EKG signal, we needed some information about how it is generated. We spent the first part of this chapter providing a brief overview of the heart's physiology and electrophysiology. We attempted to correlate the different components of the ECG signal with the various stages of a heartbeat. We then reviewed the various components of the EKG signal and discussed how different pathologies affect the shape of that component of the EKG trace. Finally, we looked at some common signal-processing tasks associated with the ECG.

The Electroencephalogram 8

The electroencephalographic signal, or EEG, is a peek into the workings of the human brain. Because obtaining the EEG is relatively inexpensive, it is a very useful diagnostic tool in low-resource environments. With the development of mobile implementations, it has the potential to provide insights into the workings of the human mind outside of laboratory conditions. To understand the EEG and use it, we need to have some basic understanding of the organ generating the signal—the brain.

8.1 The Brain

The brain is the most complex entity known to humans. With about 86 billion neurons, which can have up to tens of trillions of connections, the intricacy and complexity of the brain are beyond comprehension. The only way we can begin to get a handle on it is by simplification and gross generalizations. So, let's proceed to do just that. We can think of the brain in "evolutionary" terms, popularized in the idea of a triune brain [48] by Paul Maclean in the 1960s. In his formulation, we can divide the brain into three parts: a lizard brain, or reptilian complex, enclosed in a paleomamalian complex, which is enclosed in the neomammalian complex. The idea is that the lizard brain handles automatic actions, the neomammalian brain handles "higher-order" tasks, such as thinking, while the paleomammalian brain handles the rest; each mainly working independently of the others. While this is still a popular way of dividing the brain (except for the independence part), it is misleading. A more accurate division would be into the cerebellum, brainstem, diencephalon, basal ganglia, and cerebral hemispheres.

K. Sayood, *Biomedical Signal Processing*, Synthesis Lectures on Biomedical Engineering,
https://doi.org/10.1007/978-3-032-08855-0_8

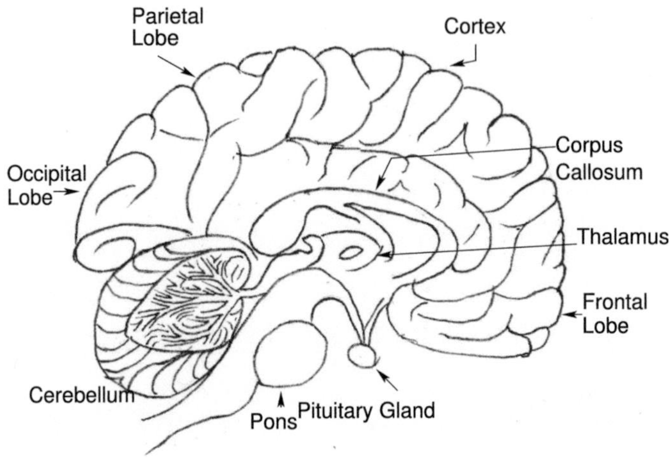

Fig. 8.1 Saggital view of a human brain

8.1.1 Brainstem

The brainstem is connected directly to the spinal cord. It consists of the *medulla oblongata* or long marrow, the *pons* or bridge, and the *midbrain*. The brainstem is involved in much of the body's automatic processing. Recall that the sinoatrial node in the heart had neural connections to increase or decrease the heart rate. The cranial nerves that do that arise in the brainstem. In fact, most of the cranial nerves arise from the brainstem. The cranial nerves play a crucial role in controlling facial activity and conveying sensory input from the face.

The medulla oblongata connects to the spinal cord and the aforementioned cranial nerve, which influences heart rate. It also contains two of the three groups of neurons involved in respiration.

The third group is in the pons, which also coordinates the activities of the cerebral hemispheres. The pons is connected to the midbrain, which contains several critical functional regions.

At the back of the midbrain are two hillock-shaped regions called the superior and inferior colliculi, or upper and lower hillocks. The inferior colliculus is a major node in the auditory pathway. The superior colliculus receives direct input from the retina and plays a crucial role in controlling eye movements. The midbrain also contains a collection of cells called the *substantia nigra*, or black substance, that is involved in the regulation of mood and motor control. Dysfunction in this region is associated with Parkinson's disease (Fig. 8.2).

The role of the brainstem is not restricted to simply handling the body's automatic processing. A very readable book that delves into the wider role of the brainstem is "The Hidden Spring: A Journey to the Source of Consciousness," by Mark Solms [49].

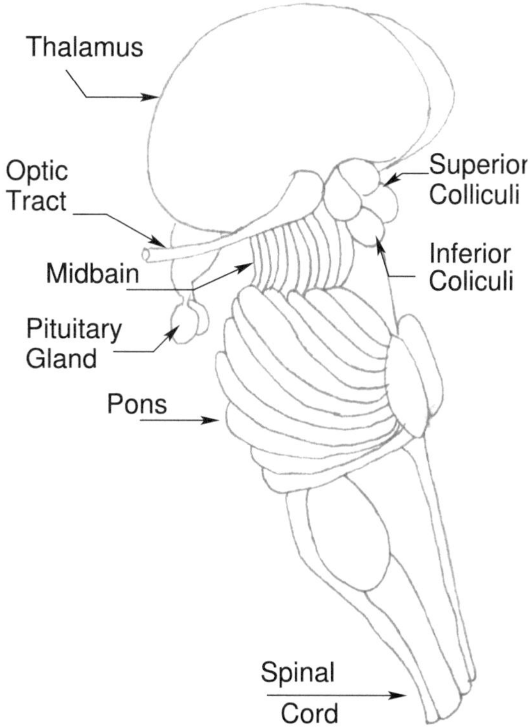

Fig. 8.2 The brainstem

8.1.2 Cerebellum

The cerebellum, Latin for "little brain," is almost a separate structure behind the brainstem. It is densely packed with neurons, with a neuron density that is significantly higher than in other parts of the brain. One way it packs this many neurons is through dense accordion-like foldings, which result in parallel grooves on its surface. The types of neurons in the cerebellum include the Purkinje cells, which are among the largest neurons, and the granular cells, which are among the smallest neurons. The Purkinje cells have numerous dendrites, which enable them to form a complex network. The cerebellum is involved in coordinating movement, maintaining posture, and regulating balance. The importance of the Purkinje cells in this function can be inferred by the fact that movement and balance can be affected when these cells are harmed for some reason. A relatively recent discovery of one of the dysfunctions is *gluten ataxia*, which is an autoimmune disease triggered by gluten in the diet. This disease is harmful to the Purkinje cells. The cerebellum is also involved in attention and language, and recent research has shown that a loss of activity in this region is associated with certain forms of autism.

8.1.3 Diencephalon

The diencephalon sits on top of the midbrain. It consists of the thalamus, or the chamber; the hypothalamus, or beneath the chamber; and the pineal, or pine tree-like, gland. The thalamus comprises two egg-shaped structures situated on top of the midbrain, joined by the interthalamic adhesion. It is a central way station, or switchboard, to the somatosensory cortex. The thalamocortical neurons project from the thalamus to many distinct parts of the cerebral cortex. It contains the lateral geniculate nucleus, which takes input from the retina and sends it to the visual cortex, and the medial geniculate nucleus, which takes input from the inferior colliculus and sends it on for auditory processing to the auditory cortex in the temporal lobe. It is also involved in regulating sleep, alertness, and consciousness. Projections from the thalamus to the cortex are reciprocated by projections from the cortex to the thalamus, establishing a feedback structure that generates some of the signals observed in the EEG. Unlike many brain regions, where inhibitory interneurons are intermixed with neuronal projections, the thalamus has a dearth of inhibitory neurons. Instead, there is a band of inhibitory neurons—a nucleus organized as a shell around the thalamus. This is the Thalamic Reticular Nucleus (TRN). The TRN sits between the neocortex and the thalamus at the intersection of the thalamocortical and corticothalamic projections. Both thalamocortical and corticothalamic projections give off collaterals that excite the TRN cells. The number of thalamocortical neurons is much less, several orders of magnitude less in humans, than corticothalamic neurons, and there are many more corticothalamic projections into the TRN than thalamocortical projections.

As the name indicates, the hypothalamus is the region below the thalamus. It involves autonomic functions such as thermoregulation, panting, and sweating, as well as emotional responses and temperature control. It also contains a group of cells called the suprachiasmatic nucleus (SCN), which is responsible for controlling circadian rhythms. The SCN receives input from cells in the retina, thus allowing it to coordinate the body's rhythms to the 24-hour cycle. The SCN controls the pituitary gland.

8.1.4 Basal Ganglia

The word ganglia is used to refer to a collection of cell bodies in the peripheral nervous system, while the word nuclei is used to refer to a collection of cells in the central nervous system, so the basal ganglia should probably be called the basal nuclei. The basal ganglia comprise a large number of structures involved in a wide range of functions, including motor control, procedural memory, habit formation, cognition, and emotion. The structures that are part of the basal ganglia include the *striatum*, which is a striped mass of grey and white matter—hence the name, the *globus pallidus* (looks like a globe and is pallid or gray), the *ventral pallidium* (more gray stuff), the *substantia nigra* or black stuff, and the *subthalamic nucleus* (group of cells below the thalamus). The striatum itself is made up of several organized collections of cells, including the *caudate nucleus* (looks like it has a tail), the *putamen* (shell-like), the accumbens nucleus (lazy bunch of cells), and the *olfactory tubercle*.

8.1.5 Cerebral Hemispheres

The cerebral hemispheres comprise the majority of the brain's mass. We can divide the cerebral hemispheres in several ways. The most obvious way is to note that there are two hemispheres: the right and the left hemispheres. The deep groove that divides them is called the *longitudinal fissure*. The two hemispheres are joined by—and communicate through—a bundle of fibers called the *corpus callosum*. Fissures or grooves or sulci (singular: sulcus) divide the cerebral hemispheres into four lobes named after the bones that cover them by the French anatomist Louis Pierre Gratiolet [50]. The central sulcus forms the boundary between the frontal lobe and the parietal lobe. The lateral sulcus forms the boundary between the frontal lobe and the temporal lobe. The parieto-occipital sulcus forms the boundary between—wait for it—the parietal and occipital lobes, and the preoccipital notch forms the boundary between the occipital and temporal lobes. All divisions shown in Fig. 8.3.

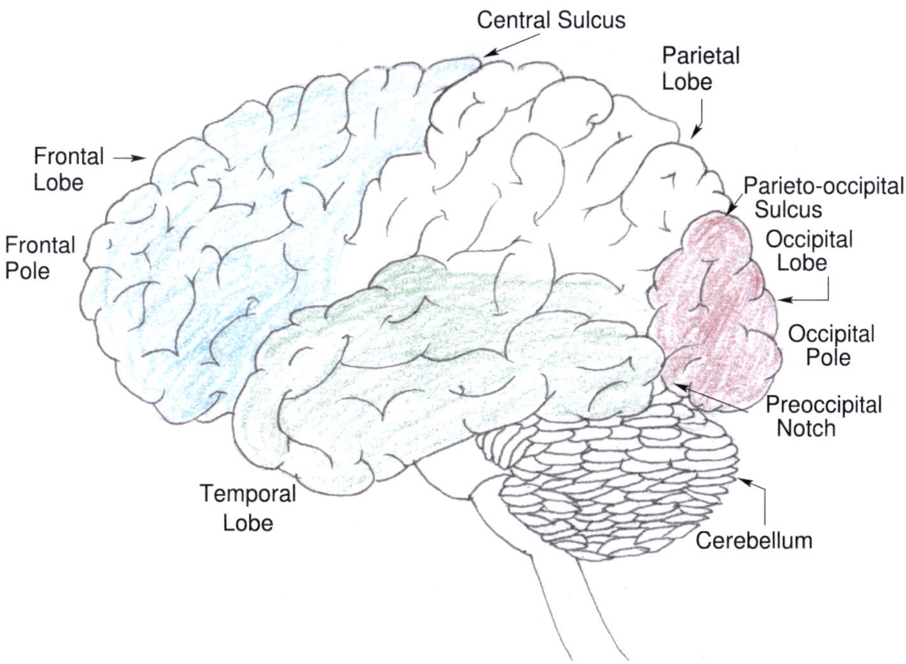

Fig. 8.3 The lobes of the cerebral hemispheres

The surface layer of the cerebral hemispheres, known as the cerebral cortex, contains approximately 16 billion neurons, which is slightly less than a fifth of the total number of neurons in the brain. It packs these neurons through a folded structure, which, when unfolded, would have an area of approximately 2.6 ft^2. The frontal cortex is divided by function into the prefrontal cortex, located at the front, and the sensory and motor cortices, located at the back. The prefrontal cortex is involved in working memory, cognition, and emotional regulation, including the regulation of mood. It is the more "intellectual" part of the brain. The posterior part of the frontal lobe, before the central gyrus, is where you find the motor cortex. The planning is in the premotor cortex, which is closer to the prefrontal cortex. The prefrontal cortex is like the executive suite; the closer you are to it, the more planning tasks you do.

The other three lobes integrate sensory information with information contained in memory. As mentioned earlier, the temporal lobes, located near the temples, receive and process auditory input from the thalamus. But that is not the only thing they do. If you have read *Do Androids Dream of Electric Sheep*? [51], you might recall the Penfield Organ. Named after Wilder Penfield, one of Princeton's lesser-known football coaches, it refers to experiments by Penfield in which he found that when he excited the temporal lobe of some of his patients with mild electric currents, he could elicit vivid memories. Thus, we know that the temporal lobe is involved in the acquisition of memory. (If you haven't read *Do Androids Dream of Electric Sheep*? you should). The temporal lobe, generally the left temporal lobe, is also the site of most of Wernicke's area, which is involved in language comprehension. Wernicke's area extends to the parietal lobe, which lies under the parietal bones of the skull. The parietal lobe also receives sensory inputs from the skin via the thalamus.

One of the most studied portions of the occipital lobe is the visual cortex, which consists of the primary visual cortex (V1), the prestriate cortex (V2), the third visual complex (V3), visual area V4, and the inferior temporal cortex (IT). The visual cortex provides one of the many examples of how the brain minimizes the bandwidth required for information transfer [52].

The primary visual cortex contains a large number of neurons containing a map of the visual field. The retinal image gets projected onto the brain in the visual area V1. The neurons in this area respond selectively to stimuli of differing orientation, spatial and temporal frequency, and color. Thus, V1 neurons provide the basic primitives that make up an object. These basic pieces are organized into increasingly more complex objects as the information is conveyed through V2 and V4 to the inferior temporal cortex. A simplified block diagram of the process of information flow is shown in Fig. 8.4. Suppose we observe a box. The information about the edges is processed in the V1 region, which is passed on to the V2 region, where the edges are organized and then passed on to the V4 region, where the information is perhaps consolidated, allowing the IT region to determine that the object is a box. Now, suppose we move our head. The box is now in a different part of the visual field, exciting different neurons in the V1 region, which then pass the information on to the V2 region and so on. This is tremendously wasteful, as there are a large number of objects in our

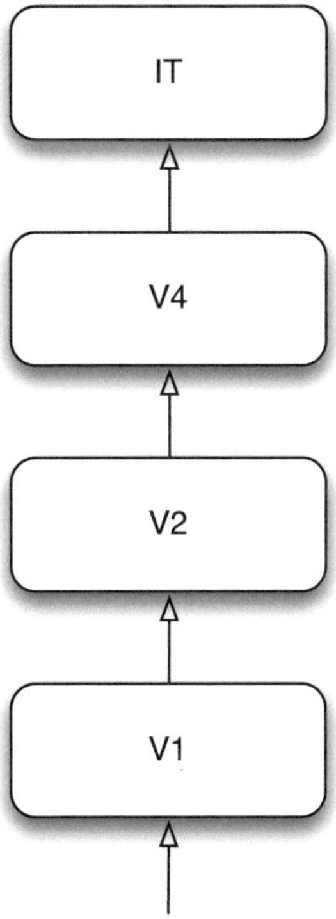

Fig. 8.4 Hierarchical processing for object recognition

visual field, and most of us tend to move our heads relatively often. If each time information processing has to start from scratch—from edges to planes to boxes—the amount of resources devoted to just image processing would have to be extraordinarily high. The same would be true for auditory and other sensory inputs. The prevailing view, supported by empirical evidence, is that each layer in the visual cortex models and predicts the input to the preceding layer. The information that gets passed on is the difference between the prediction and the input. The model construction has both a long-term and a short-term component with input from many parts of the brain. A possible simplified diagram of the information flow is shown in Fig. 8.5. This looks a lot more complicated than Fig. 8.4, however, this complexity allows for a significant reduction in the need for information processing. Information from other parts of the brain and the state of each level is used to generate a prediction of the possible inputs to a lower level. Information flow from the lower level to the upper level occurs only when the prediction based on the model at the upper level differs from the input to the lower level. Therefore, if the model in the upper level is accurate, no information flows into the upper level, obviating the need for processing. If, on the other hand, the model is inaccurate, information about the error in prediction (commonly known as prediction error) flows from the lower level to the upper level, inducing an updating of the model. As the world, in general, does not change in surprising ways, this considerably reduces the amount

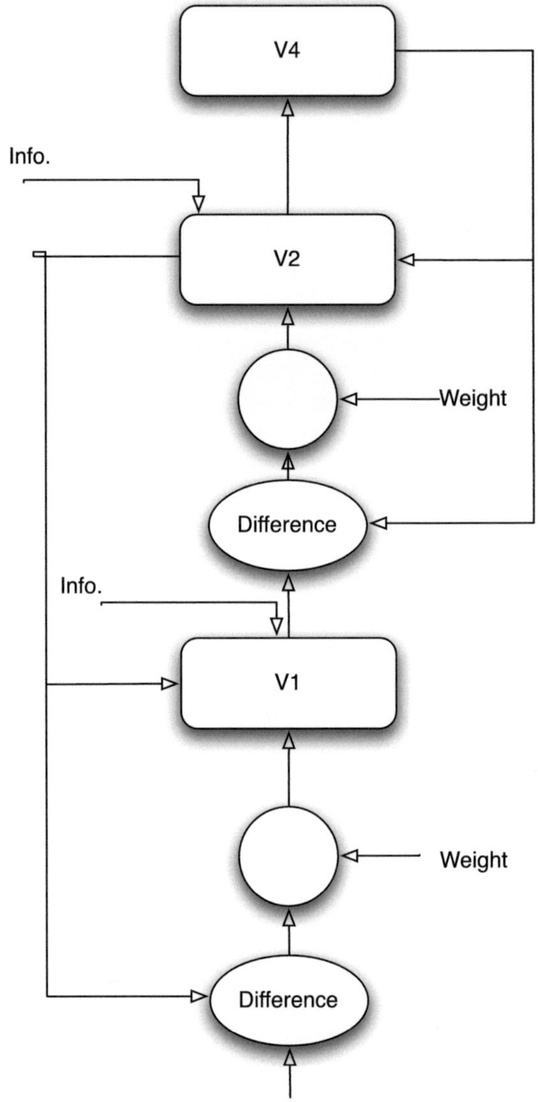

Fig. 8.5 A prediction error model of visual perception

of information that needs to be processed as we go merrily turning our heads to view objects. Each time we turn our head, the information about the motion of the head and the previous location of the box would allow for quite accurate predictions about the composition of the visual field, resulting in minimal stimulus proceeding up the forward path. In our block diagram in Fig. 8.5, we have also included a “weight" input. This, in some sense, represents the attention we are paying to our perception. If very little attention is paid, the weight given to the prediction error will be small, resulting in the acceptance of the model even when it does not quite align with the perceptual input.

8.2 The Electroencephalograph

8.2.1 History

The discovery of EEG is generally credited to Hans Berger (1873–1941). He recorded the first human EEG (his son's) in 1924 and developed and named the electroencephalogram in 1926. Berger was an officer in the Prussian army when, after a traumatic fall from his horse, he received a letter from his sister who had dreamt that he had a terrible accident. Convinced of the existence of telepathy, he focused his later studies on the electrical activity in the brain [53]. While Berger justifiably gets much of the credit, as with all great discoveries, his work was simply the bow that tied up the package that is the electroencephalogram. Richard Caton demonstrated the variations in electrical potential in the exposed brains of rabbits and monkeys in 1876. In 1890, Adolf Beck examined the electrical activity in the brains of dogs when excited by spots of light. Before them, Ernst Fleischl von Marxow (1846–1891) demonstrated that sensory stimulation led to changes in electric potential in various regions of the skull. He didn't publish his results, though, preferring to bury them in a vault. He did gain considerable fame as the victim of his admirer, Sigmund Freud's medical malpractice. Fleischl von Marxow had to have an infected thumb amputated and, to deal with the resulting pain, had become addicted to morphine and heroin. Freud prescribed cocaine to cure his addiction and claimed the treatment had been phenomenally successful (spoiler alert: it hadn't). He even wrote a paper about it called, appropriately enough, "on Coca." Fleischl succumbed to the tender ministrations of Dr. Freud and died at the age of 45. Hopefully, he died happy.

8.2.2 What Does the Electroencephalogram Measure and What It Cannot Measure

As you might recall, the brain contains a dense collection of billions and billions of neurons, which are heavily interconnected. As you may also recall, these neurons communicate with each other through action potentials. It makes sense, then, that there would be electrical activity across the brain. How much electrical activity can be deduced by the fact that under three pounds, the brain is only a small fraction of the body mass, yet it consumes more than 20 percent of the energy used by the body. It uses more than half of the total blood glucose in the body. This energy is being used to generate electrical activity, which can be detected by placing electrodes on the skull. However, because the brain is wrapped up in multiple layers of covering—the pia mater, the arachnoid mater, and the dura mater—and the bony skull, not to mention the layers on top of the skull, the signal obtained via the electrodes on the skull has severe limitations in terms of information transmission. Each electrode placed on the skull receives about half of its signal amplitude from directly below the electrode and about 95% from within a 6 cm radius.

Trying to understand exactly what is going on in the brain, electrically speaking, using the EEG would be like placing a small number of microphones on the roof of the Astrodome and trying to figure out the conversations taking place in the stands. Furthermore, the EEG signal is disrupted by muscular activity in the skull. For example, eye blinks can cause significant distortions in the output of the leads close to the eyes. A considerable amount of signal processing is required for the EEG signal to be a useful reflection of activity in the brain.

Despite all these problems and despite repeated predictions of its demise, the EEG has survived. Most of these predictions of the demise of the EEG relate to the increasing availability of high spatial resolution imaging technology such as Magnetic Resonance Imaging (MRI) and Magnetoencephalography (MEG). The EEG has several advantages over these techniques, which should keep it viable for the foreseeable future. The EEG is significantly cheaper to obtain than any of the new imaging techniques. Unlike MRI or MEG, which require massive, fixed installations, the instrumentation needed for obtaining EEGs is light and highly portable. Especially with the arrival of dry electrodes, the EEG can be used in situations where the subject has to move around. Patients can generally tolerate the electrodes required for EEG recordings. The process is quiet, and the subjects are not exposed to potentially harmful radiation.

8.2.3 The 10–20 System

Skulls come in a variety of sizes and shapes. To accommodate this variety while maintaining a degree of standardization, we require a standardized system for placing electrodes that can accommodate variations in size and shape. This standard system is called the 10–20 system.

To explain this system, let's start with some landmarks. The groove behind the bridge of the nose is called the *nasion*. The little bony protrusion at the back of the head is called the *inion*. The other two landmarks are the pre-auricular points. These points are small protrusions just in front of the ear. You can feel them if you put your finger in front of your ear and open and close your mouth. Let's draw a circle to represent the top view of the skull and mark off the landmarks as shown in Fig. 8.6. We now draw a line between the nasion and the inion—this line is called the midline—and partition it as follows: We mark off two points, each covering 10% of the distance between the nasion and the inion, at each end of the midline. The remaining part of the midline is broken up into four equal intervals, each covering 20% of the distance between the nasion and the inion. This division of the midline is the reason for the name of the 10–20 system. We do the same for the line joining the two pre-auricular landmarks. Now draw a circle centered at the center of our idealized view

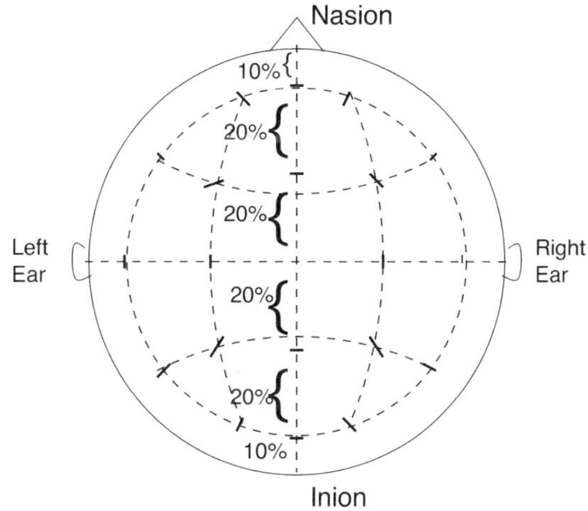

Fig. 8.6 Guidelines for placing electrodes according to the 10–20 standard

of the skull and with a radius equal to 40% of the midline. Starting at the intersection of the horizontal line connecting the two pre-auricular landmarks, mark off ten equal-length segments. As shown in Fig. 8.6, draw arcs connecting the two marked-off points on both sides of the nasion to the corresponding points on the two sides of the inion. These arcs connect the marked-off points on the circle and the 20% points on both sides of the midpoint of the circle. Do the same for the pair of points on both sides of the center line.

If we now mark the points where the guidelines intersect and then delete the guidelines, we obtain the points for electrode placement according to the 10–20 standard, as shown in Fig. 8.7.

We can now place electrodes at these locations. The names of the electrodes depend on their location with respect to the lobes of the cerebral cortex, the hemisphere, and their proximity to the midline. The first letter denotes the lobe of the corresponding cerebral cortex. Electrodes placed over the frontal lobe are named with the letter F, the electrodes close to the parietal lobe have the letter P as the first letter of their name. Similarly, electrodes close to the temporal lobe begin with the letter T, and the electrodes over the occipital lobe begin with the letter O. In the case of electrodes close to the prefrontal cortex and those on the midline, we have an additional letter in their names. The names of the electrodes situated on the line joining the two pre-auricular landmarks that are not over the temporal regions begin with the letter C. The names of the electrodes close to the prefrontal cortex begin with Fp, and the names of the electrodes on the midline have a lowercase z as the second letter in their name. Except for the electrodes on the midline, the names of all other electrodes end with a number. The convention is to assign odd numbers to the electrodes in the left hemisphere and even numbers to the electrodes in the right hemisphere. Lower numbers are assigned to the electrodes closer to the midline (Fig. 8.8).

Fig. 8.7 Possible locations for placing electrodes

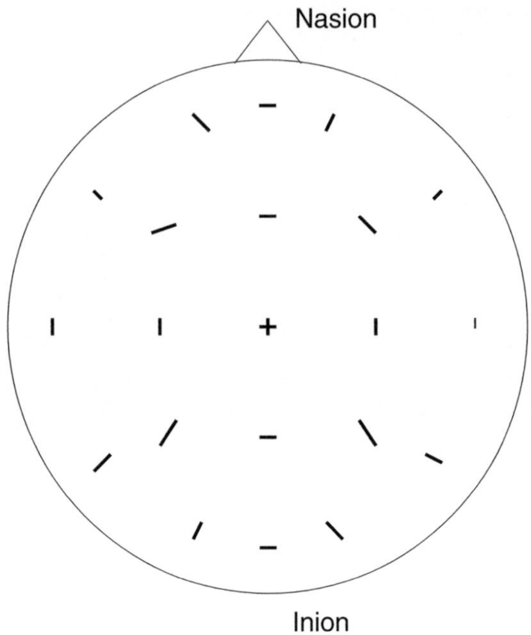

Fig. 8.8 Electrodes placed according to the 10–20 standard

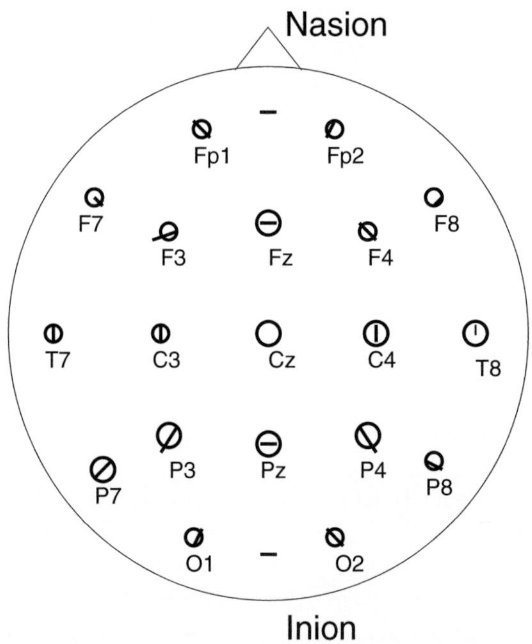

8.2.4 Montages

The electrodes of the EEG are points of contact with the skull. To determine voltages, we need two points. How we select these points is described in the *montage* we use. Montage seems like a strange word to use for describing how we obtain voltages, so let's briefly describe why we use this word. According to the Merriam-Webster dictionary, the word montage is defined as:

1. the production of a rapid succession of images in a motion picture to illustrate an association of ideas
2. (a) a literary, musical, or artistic composite of juxtaposed more or less heterogeneous elements
   (b) a composite picture made by combining several separate pictures
3. a heterogeneous mixture

We use the word montage because of the way the various voltage signals are displayed. Look at the example in Fig. 8.9. Clearly, this fits definition 2(b) of montage.

8.2.4.1 Bipolar Montage

The example shown in Fig. 8.9 is an example of a bipolar montage. In this particular montage, the voltage signal displayed is the voltage difference between two consecutive electrodes in a link, which is part of a chain of sequential electrodes. The most popular bipolar montage is the double banana bipolar montage. An example of this is shown in Fig. 8.10. The chain in this montage on the outside right goes from an electrode in the prefrontal region, $Fp2$, to an electrode in the frontal region, $F8$, to an electrode in the temporal region, $T8$, to the outside electrode in the parietal region, $P8$, and finally to an electrode in the occipital region $O2$. This chain is referred to as the *right temporal chain*. Another chain also begins with $Fp2$ and terminates on $O2$, but through the electrodes interior to the ones in the right temporal chain. This chain is referred to as the *right parasagittal chain*. The voltages displayed for the right temporal chain will be the voltages between $Fp2$ and $F8$, $F8$ and $T8$, $T8$ and $P8$, and between $P8$ and $O2$, with the first electrode in each pair assumed to be at a higher potential. There is a similar arrangement on the left side of the skull. You can see why the montage is called a double banana montage. Besides these two chains, there is also a *midline chain* from Fz to Cz to Oz. The signals in this montage are typically displayed with all the voltages from the temporal chains at the top, followed by the signals from the parasagittal chain, and then any other voltages, such as those from the midline chain.

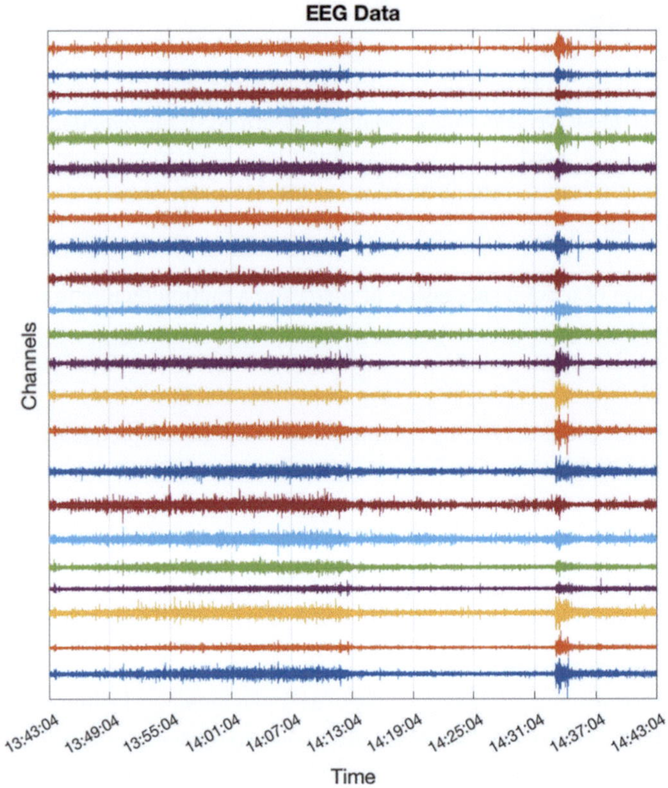

Fig. 8.9 An example of a bipolar montage

8.2.4.2 Referential Montage

We can also create a montage of signals where each electrode is referenced to a common reference point. This type of montage is called a referential montage. Common references used include the auricular points by the ears, the Cz electrode, or the average value of all the electrodes.

8.3 Preprocessing

Preprocessing of the EEG signal involves removing extraneous signals. The EEG signal itself is assumed to lie between 0.01 and 100 Hz. Within this range, the signal is partitioned into different frequency bands, each with somewhat different localization. The bands are

delta: These are the frequencies below 4 Hz. They generally appear in the frontal region in adults. In slow wave or deep sleep, delta waves are the dominant components of the EEG

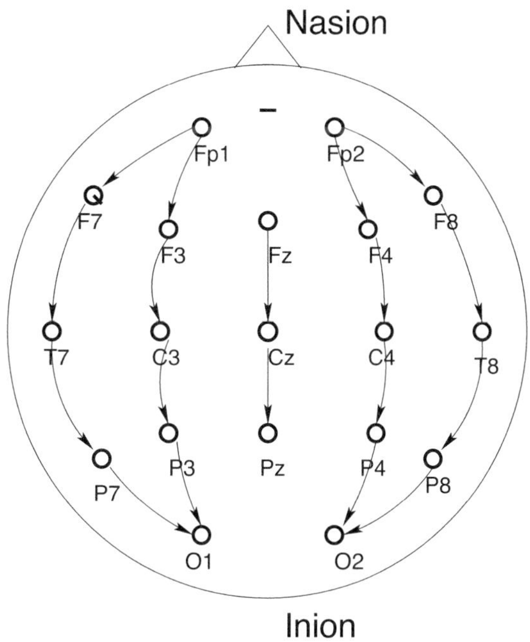

Fig. 8.10 An example of the configuration of electrodes in a double banana bipolar montage

signal. As babies spend a significant amount of time in slow-wave sleep, they exhibit a strong delta component in their EEG signals.

theta: These are the frequencies between 4 Hz and 7 or 8 Hz—different sources use different conventions. The occurrence of these waves is indicative of deep relaxation and usually appears when the subject is drifting in or out of sleep. Diffuse theta waves are also seen after a generalized seizure.

alpha: These are the frequencies between 8 and 12 Hz. These frequencies are primarily observed in the occipital region and occur when the subject is relaxed with their eyes closed. It seems that the name alpha should have been assigned to the lowest frequencies—the frequencies that are called delta, rather than the third band of frequencies. The reason for this apparent anomaly is that these were the waves Hans Berger first saw with his son. Hence the name "alpha."

beta: The component of the EEG signal between 12 to 30 Hz is called the Beta component of the EEG signal. These frequencies generally show up in the frontal and parietal regions and are associated with heightened states of awareness, including cognition, and working memory tasks.

gamma: The Gamma frequencies are variously defined as the frequencies above 30 Hz or the frequencies in the 30–100 Hz region. These higher frequencies typically appear when people are engaged in activities that involve a large portion of the brain. These include working memory and other cognitive tasks. According to one of the more popular theories of consciousness, the Global Neuronal Workspace Theory [54, 55], an extension of the Global Workspace Theory [56], consciousness involves the recruitment of large areas of the brain. In a way, then, the gamma frequencies reflect the fact of consciousness.

For biological signals, this is a very wide range of frequencies, one that is also occupied by numerous other signals. The line frequency of 60 or 50 Hz, depending on the geographic location, lies right in the middle of the Gamma frequency range. The (hopefully) omnipresent EKG signal occupies a significant portion of the lower frequency range. The movement of the eye and eye blinks especially affect signals from the frontal electrodes. Finally, there is the electrical signal generated by the muscle cells' activities, known as the electromyogram or EMG. Therefore, for the EEG signal to be useful, preprocessing is a necessity. The preprocessing falls under the general heading of artifact removal.

8.3.1 Artifact Removal

The power line signal and the EKG signal are ubiquitous, so the best way to reduce their effect is to use a differential amplifier with a high common-mode rejection ratio (CMRR). However, this only reduces the effect, not eliminates it. Let's briefly examine the various methods used to remove or mitigate the effects of different artifacts.

8.3.1.1 Filtering

Filtering is most helpful in removing two artifacts—baseline wander and line noise. To remove baseline wander, as in the case of the EKG, we use a high-pass filter. Generally, a very low-order filter is both sufficient and advisable. The easiest thing to do is design a filter by putting a zero at the location $z = 1$ in the z-plane and then putting a pole or a pair of complex conjugate poles to cancel out the effect of the zero.

To eliminate line noise, we generally use a notch filter at 60 or 50 Hz. This does impact the gamma band; however, the signal loss seems to be generally acceptable. '

8.3.1.2 Linear Regression

The eye blink and eye movement signals can be removed by using signals obtained from electrodes positioned next to the eye to generate an electrooculogram (EOG). There is a potential difference between the cornea and the retina. When the eye moves towards one of the electrodes, this causes a change in potential, which is reflected in the EOG. A change in potential is also present during eye blinks. We can use the EOG signal to remove the eye-blink and eye-movement artifacts from the EEG signal, the idea being that the EEG signal being recorded is a sum of the "true" EEG signals and the EOG signal. Therefore, we can simply subtract the contribution of the EOG signal from the measured EEG signal to obtain the "true" EEG signal. There is an inherent assumption that the EOG signal itself is uncontaminated, which is not entirely accurate. In practice, the EOG signal is also contaminated by the EEG signal.

8.3.1.3 Time-Frequency Decomposition

The EEG signal is a composite signal. In the simplest model, it is the sum of different frequencies. In Chapter 3, we demonstrated how the FFT can be used to extract the different frequency components from a composite signal. Consider a signal that is the sum of three sinusoids at frequencies 1 Hz, 27 Hz, and 42 Hz.

$$f(t) = \cos(2\pi t) + 1.5\cos(54\pi t) + \cos(84\pi t)$$

A plot of the signal and the corresponding plot of the magnitude of the FFT components is shown in Fig. 8.11. In the frequency picture, we can clearly see the three components with the same relative magnitude as the original signal. Extracting any of the components from the Fourier coefficients would be easy.

However, there is an underlying assumption—valid in this case—that the components are present at all time instants. In most EEG signals, this is not the case. Various frequencies turn on and off. An idealized situation with the same three frequencies is one where the 1 Hz signal is always present, the 27 Hz signal appears at t = 0.1 s and disappears at t = 0.3 s, and the 42 Hz signal appears at t = 0.5 s. The plots of the time and frequency domain

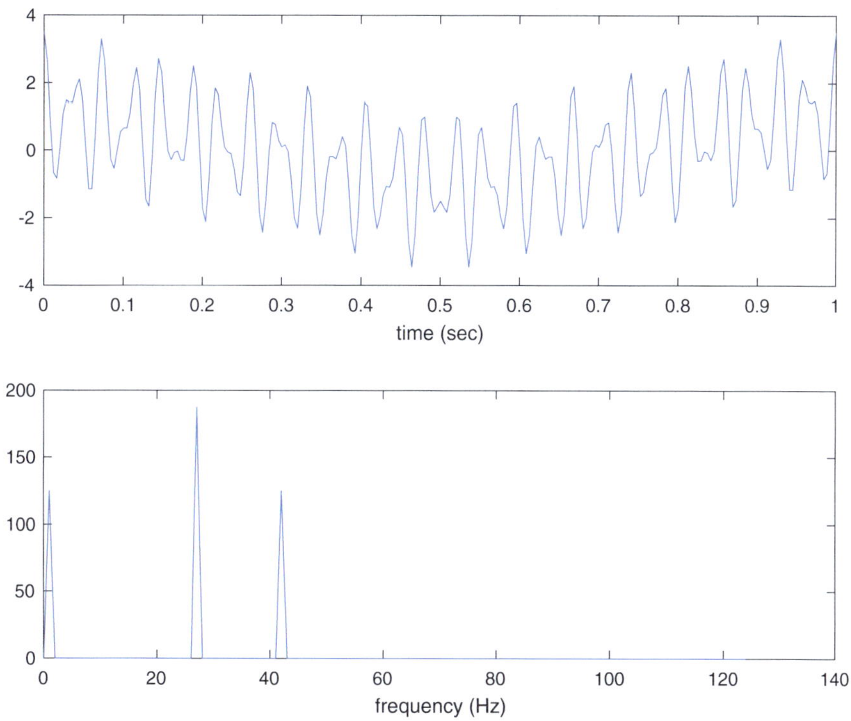

Fig. 8.11 The time and frequency pictures of a sum of sinusoids

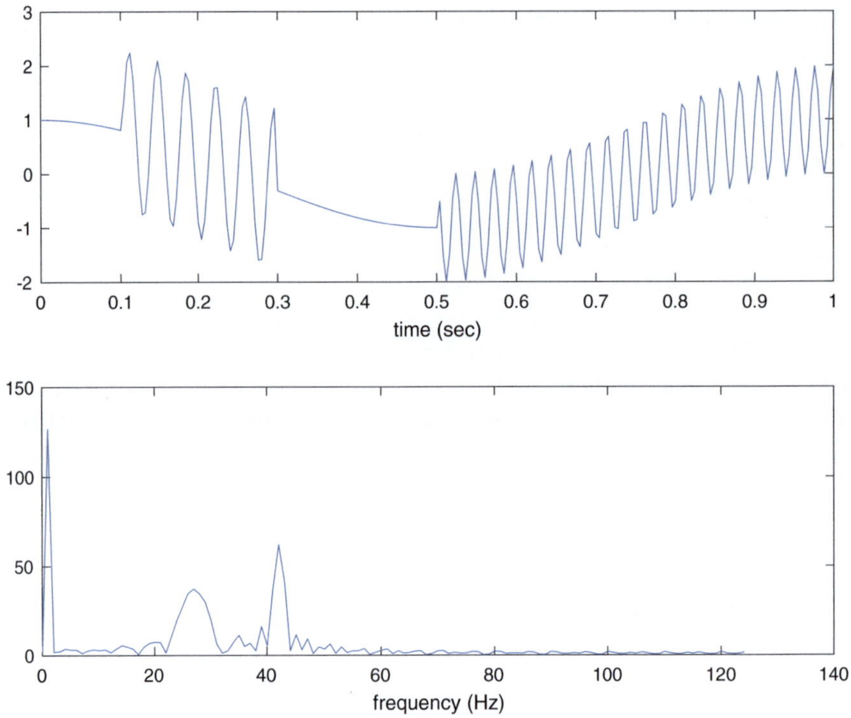

Fig. 8.12 The time and frequency pictures of a sum of sinusoids

signals are shown in Fig. 8.12. Notice that one piece of information that is missing from this frequency picture is the time localization of the different frequency components. We don't know exactly when the 27 Hz signal turned on and off, and we can't tell when the 42 Hz signal turned on and off. With some thought, we can tell that these signals were turned on and off during that one-second interval. The turning on and off of a signal can be modeled as a product of a signal that exists for all time and a signal that is zero except during the interval when the signal is on, during which time its value is one. An example is shown in Fig. 8.13.

The Fourier transform of a product is the convolution of the Fourier transform of the factors in the product. Therefore, the Fourier transform of the sinusoid that turns on and off is the convolution of the impulse functions, which are the Fourier transform of the sinusoid, and a Sinc function, which is the Fourier transform of the pulse. Hence, the smearing out of the Fourier transforms of the sinusoids at 27 and 42 Hz. Notice that the Fourier transform of the 1 Hz sinusoid is not smeared, as the 1 Hz signal exists for the entire second.

Another piece of information is the relative magnitude of the three components. This is again because, over the one-second time interval, the amount of power in the 1 Hz signal is greater than the amount of power in the 42 Hz signal, as the 1 Hz signal lasts for the entire

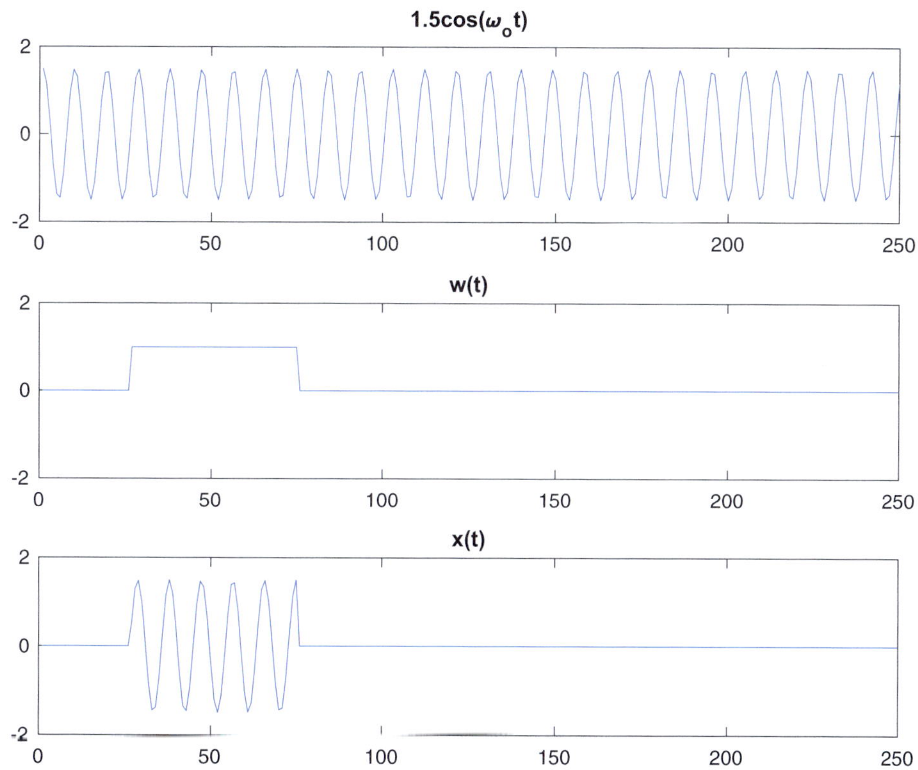

Fig. 8.13 A sinusoid that turns on at t = 0.1 and turns off at t = 0.3 can be modeled as a sinusoid that exists for all time multiplied by a pulse that is one between t=0.1 and t=0.3 and zero everywhere else

second while the 42 Hz signal lasts for only half a second. Similarly, the 27 Hz signal lasts for only a fifth of a second, so its magnitude is the smallest in the frequency picture.

So, what do we do in these situations? One way to obtain a picture of the signal in both time and frequency is by taking the Fourier transform over small intervals that cover the period of interest. In our example, instead of taking the Fourier transform of the entire second worth of data, we could take ten Fourier transforms, each over 0.1 s of data. If we do that, we get ten different Fourier transforms whose magnitudes are plotted in Fig. 8.14. This kind of analysis, which is a function of both time and frequency, is often referred to as time-frequency analysis.

You can see from the figure that neither the 27 Hz signal nor the 42 Hz signal shows up in the 0–0.1 time window or the 0.3–0.5 time window. The 27 Hz signal only appears in the 0.1–0.3 window, and the 42 Hz signal only appears in the windows starting from 0.5 s. This localization agrees with the time picture.

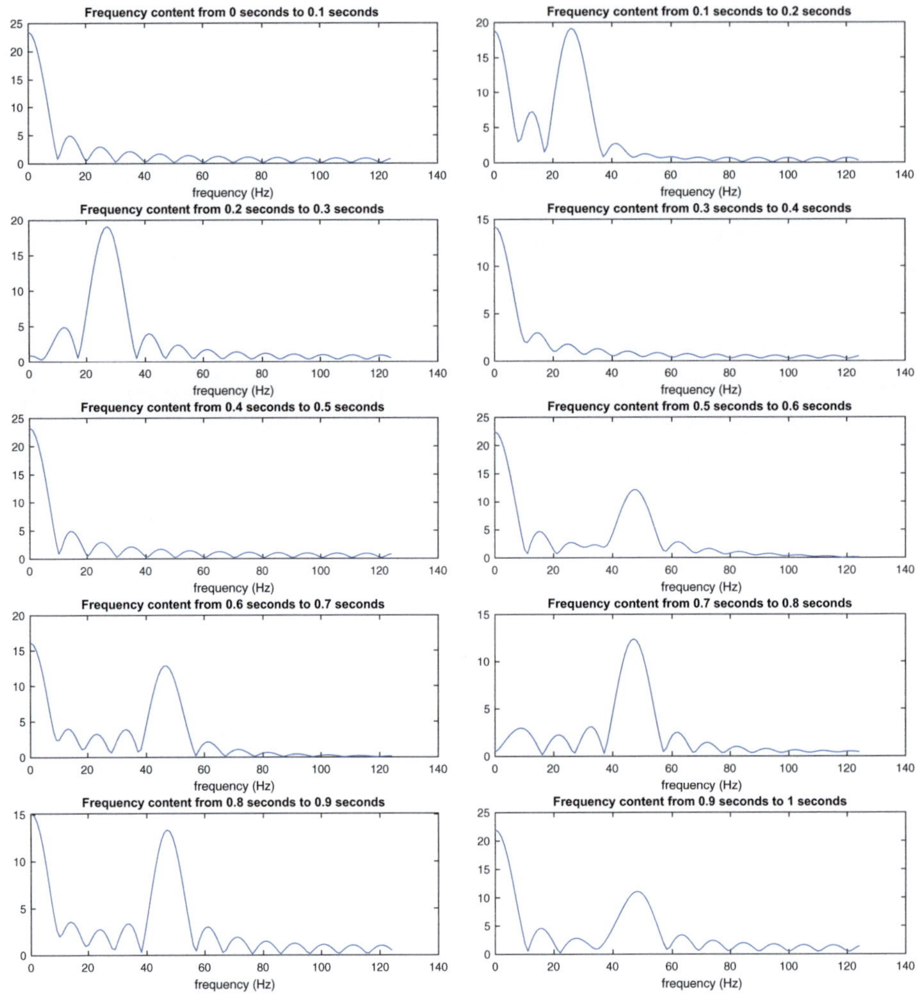

Fig. 8.14 A time-frequency analysis of the example waveform

The Short-term Fourier Transforms are essentially the Fourier transforms of the product of the original signal with a shifting rectangular pulse. Hence, the ripples in the transform. We can improve the picture a bit by using a Hanning window instead of a rectangular window, as shown in Fig. 8.15.

Notice the 1 Hz signal does not appear in any of the windows. This is because, to capture a particular frequency, you need to examine the signal over at least one period. A window of one-tenth of a second is not sufficient to capture a signal whose period is one second. This is a problem. To capture low-frequency signals, we need windows of longer duration. In this case, we would have needed a window that is at least one second long. However, as

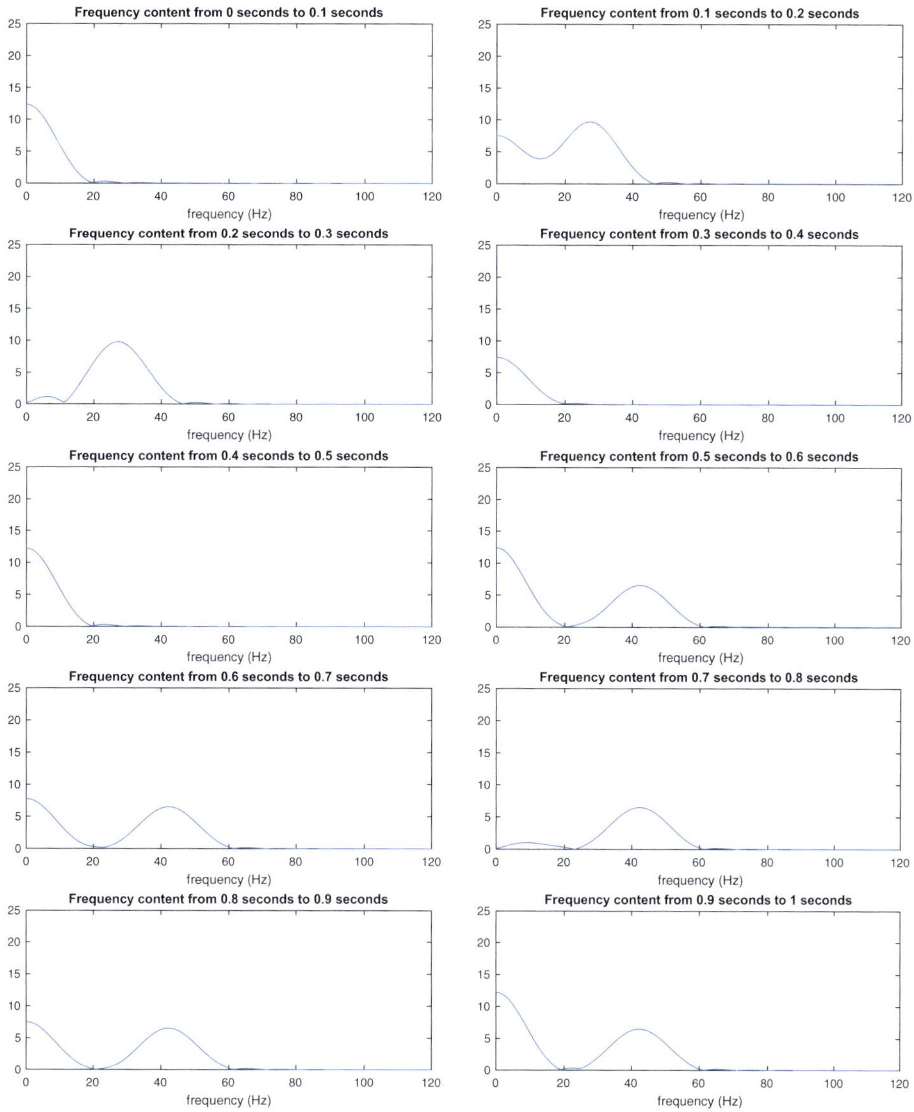

Fig. 8.15 A time-frequency analysis of the example waveform using Hanning windows instead of rectangular windows

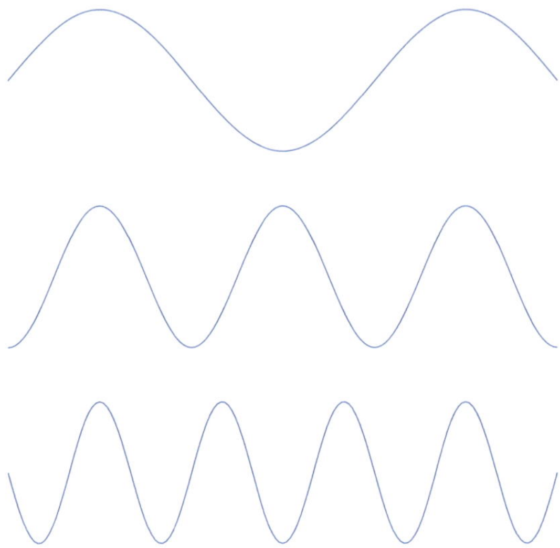

Fig. 8.16 Three basis functions for a short-term Fourier transform

we have already seen, a window one second long does not allow us to localize the 27 Hz and 42 Hz signals. To localize these, we need windows of short duration. Enter wavelets. Wavelets are used for time-frequency analysis, employing longer windows to extract the lower frequencies and shorter windows to extract the higher frequencies. To illustrate our point, let's examine an example [15].

Recall that when we take the Fourier transform, we are rewriting the function being transformed in terms of a basis set made up of sines and cosines. For example, in Fig. 8.16, we can see the first three basis functions for a short-term Fourier transform. The window over which they are non-zero is the same in all three cases, but the frequency of the basis vectors increases. As the window size remains the same, the number of periods contained within the window increases with frequency. For the wavelet case, we might have the three basis functions shown in Fig. 8.17. In this case, the window size and the frequency are related. While the number of periods for each basis function in this case is the same, the lower frequency basis function occupies a larger window, while the higher frequency function occupies a smaller window. We get higher frequency resolution at lower frequencies and higher time resolution at higher frequencies. This is a somewhat crude example. Several resources, such as [15, 57], can provide a more comprehensive picture if you are interested in using wavelets. If you are truly interested in this subject, there is the classic work by Ingrid Daubechies, *Ten Lectures on Wavelets* [58].

Most development environments offer functions that enable the use of wavelets for time-frequency analysis. MATLAB has an entire toolbox devoted to wavelets. An introduction to some of the popularly used functions in MATLAB can be found in [59]. In the discrete domain, where most biomedical signal processing occurs, wavelet functions are used to obtain a bank of filters for time-frequency analysis. Commonly used wavelet families include

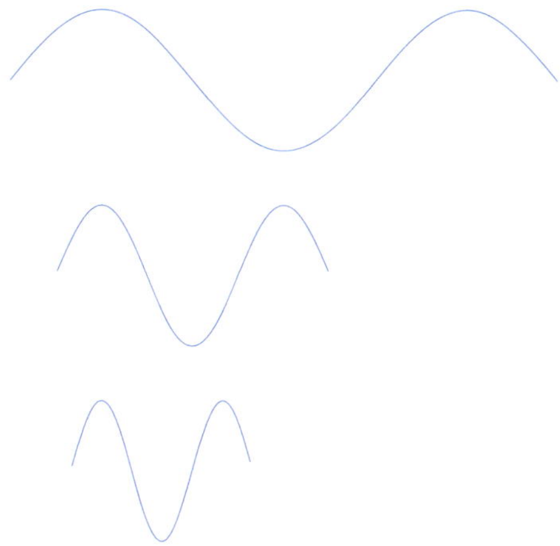

Fig. 8.17 Three basis functions for wavelet decomposition

Daubechies, *Coiflet*, *Haar*, and *Morlet*. MATLAB functions for discrete time-frequency analysis of signals include *wavedec* and *modwt*. Let's use *modwt* with our problematic signal. If x is our signal *modwt(x,'db4',3)* will return the outputs of four overlapping filters. These are shown in Fig. 8.18, where the first plot is the plot of the data, followed by the output from the highest frequency filter, all the way down to the lowest frequency.

Notice that we obtain the 1 Hz signal as the output of the low-pass filter, Filter 1—something we couldn't do with the short-term Fourier transform. The 27 and 42 Hz signals are the outputs of Filter 2 and Filter 4, respectively. In each case, there is some contamination from the other signals.

8.3.2 Feature Selection

The features selected from the EEG signal are often application-dependent. We will take a closer look at these when we examine the particular application. However, several aspects of the EEG signal are widely utilized as features in various applications. One of them is the role of different frequency components of the EEG signal.

8.4 Detecting Changes in Brain States

The brain is a highly dynamic and complex system. In this context, the EEG is an extremely crude instrument. However, some changes in the brain are so extreme that even the EEG can detect them if the EEG signals are processed properly. This happens when a person suffers an epileptic seizure or a concussion, and it also happens during sleep. There are also situations where communication between the brain and the rest of the body, or between the brain and

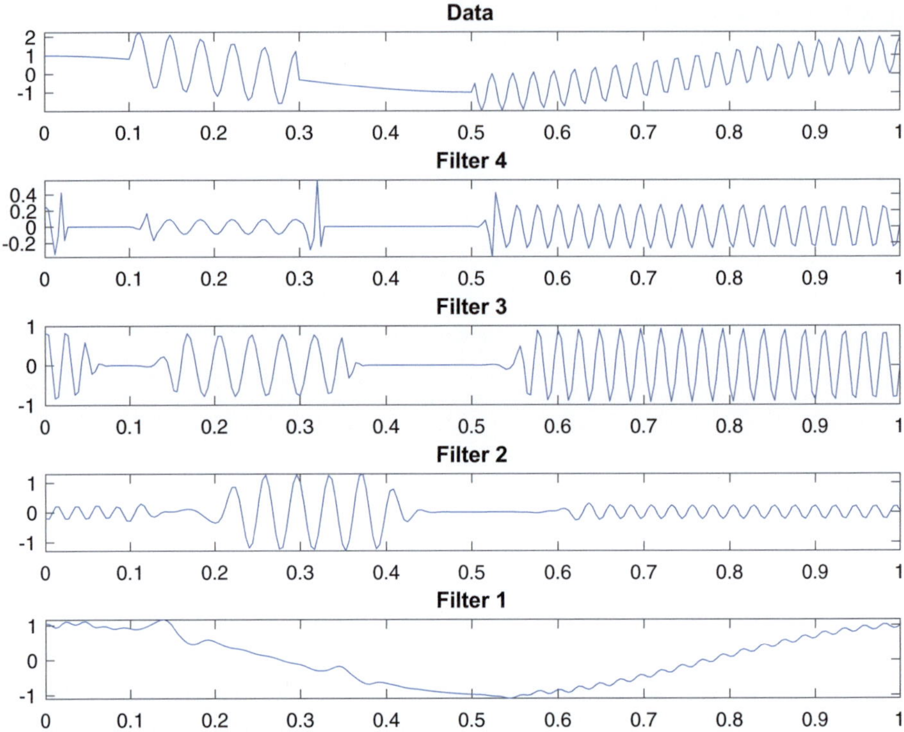

Fig. 8.18 Original data and outputs of four filters using the eight-tap Daubechies filters

the external world, is disrupted, and we require alternative means of communication. Let's examine how EEG is utilized in these situations.

8.4.1 Epileptic Seizures

A seizure is a condition characterized by a significant amount of synchronous neuronal activity in the brain. It can be reflected in convulsions, such as in the case of what used to be called grand mal seizures and is now called generalized tonic-clonic seizures (GTCS). While this is the movie view of seizures, quite often, seizures can occur without any obvious muscular convulsions. Seizures can be attributed to specific causes, such as low blood sugar or low sodium, or their onset may not be attributable to a particular cause. The former types of seizures are referred to as unprovoked seizures, while the latter are referred to as provoked seizures. Epilepsy is a neurological condition that results in recurring unprovoked seizures. The International League Against Epilepsy (ILAE) defines epilepsy as:

1. At least two unprovoked (or reflex) seizures occurring more than 24 h apart.
2. One unprovoked (or reflex) seizure and a probability of further seizures similar to the general recurrence risk (at least 60 %) after two unprovoked seizures, occurring over the next 10 years;
3. Diagnosis of an epilepsy syndrome.

This somewhat fuzzy definition reflects the fact that Epilepsy, while classified as a disease, is actually a collection of disorders that share the common symptom of a seizure. Most epilepsy patients in the world do not get treated, in large part because their seizures do not get detected. For a seizure to be detected, it must occur in a manner that can be observed by a person who is able to recognize it. As we have mentioned, there are many kinds of seizures, perhaps the majority, that are difficult to detect. Sometimes, even the person having the seizure is not aware of the occurrence. Due to this, seizure detection is a crucial area of research in biomedical signal processing. Because seizures can occur infrequently, knowing when a seizure is likely to occur could significantly enhance the quality of life for many Epilepsy patients. Seizure prediction is another major area of research in EEG signal processing. Let's take a brief look at the area of seizure detection.

8.4.2 Seizure Detection

While grand mal seizures are easy to detect and identify simply by direct observation, this is not the case with most epileptic seizures. Many seizures can be difficult to detect for non-specialists, even with the availability of EEG signals. Thus, many epileptic patients go untreated. This is tragic because the treatment for the vast majority of epileptic conditions is simple and cheap. One major problem with developing algorithms for seizure detection is the complex nature of the brain. The complexity of the brain also allows it to be highly individual, meaning that seizures in different people may manifest differently on the EEG. Another is the variety of seizure types. The seizure can have a focal onset; that is, it begins in one hemisphere and may be localized to a particular region of the brain. It can have a generalized onset, in which the seizure affects both sides of the brain simultaneously. The person can be aware of the seizure, or the awareness may be impaired. The seizure can result in loss of muscle tone or symmetrical or asymmetrical jerking of the same group of muscles. This variety means that, with a crude instrument like an EEG, the manifestation of a seizure can differ significantly among individuals and for different types of seizures. As a result, many seizure detection algorithms are either patient-specific or seizure-specific.

The seizure detection problem can be viewed as a binary feature classification problem. Recall from Chaps. 4 and 5 that to solve a feature classification problem, we need to define a set of features and select a suitable feature classification algorithm. For example, consider the algorithm proposed by Shoeb et al. [60]. In this algorithm, the features used for classification characterize the morphology and the spatial localization of the EEG waveform. To extract the features characterizing the morphology, the algorithm uses time-frequency analysis. The EEG signals are divided into epochs of two seconds each and filtered using filters based on the Daubechies wavelet family. The magnitude of the output of the four filters, corresponding to frequencies in the range of 0.5 Hz to 25 Hz, is then computed, which yields four features. If $x_{i,j}[k]$ is the kth output of the jth filter operating on the ith EEG channel then these features are given by

$$f_{i,j} = \log\left(\sum_k |x_{i,j}[k]|\right), \quad j = 1, \ldots, 4, \; i = 1, \ldots, 21$$

For each lead, a feature vector of size four is constructed. In this paper, the authors report two distinct methods for building the classifier. In the first case, all these feature vectors were put together into a single 84-dimensional feature vector. This was then classified as a seizure or non-seizure event using a support vector machine that had been trained on the same patient. In the second case, the feature vectors from each channel are fed into separate support vector machines, each trained on the same channel from the same patient. The channels are organized into different groups based on the spatial proximity of the corresponding leads. These groups, in effect, represent the spatial localization feature of the algorithm. During the training session, the group showing the greatest separation between seizure and non-seizure epochs is selected. During the test phase, if more than half of the support vector machines in the selected group declare a seizure is present for more than 6 s, the algorithm declares a seizure.

This approach, which involves using time-frequency analysis to extract features and then employing classification algorithms to distinguish epochs as either reflective of a seizure or a nonseizure, has become quite popular. There are many variations in terms of the time-frequency analysis and the classification. We can employ various wavelet-based approaches for time-frequency analysis, as well as different classification methods, including support vector machines with diverse kernel functions, the generalized Lloyd algorithm, and different types of neural networks. Beyond this basic structure, one turns to aspects of the particular form of the EEG waveform during seizures to refine this basic structure. A review of some of the recently proposed detection algorithms can be found in [61].

Epileptic seizures are very significant events in the brain. Therefore, despite the great variability in both patients and types, it is possible to develop approaches that are patient-agnostic. An example of this approach is the algorithm proposed by Mansouri et al. [62], which leverages the fact that during a seizure, there is a significant increase in synchronized activity. As with most seizure detection methods, they begin with time-frequency analysis. The EEG signal is divided into 10-s epochs with a 5-s overlap between epochs. For the frequency decomposition, instead of wavelets, they use short-term Fourier transforms. Seizure detection is performed in two steps. In the first step, an epoch is evaluated to determine whether it could be a candidate for seizure. The power in the theta-alpha band is used for this purpose. During a seizure, the power in this band goes up significantly. However, the power in this band can also go up for other reasons. For example, the power in this band increases when the person is engaged in cognitive tasks. Generally, though, when transitioning into a seizure, the power tends to go up rapidly. This means that, unlike during cognitive tasks, the power in these bands during the epochs preceding the seizure epoch will be relatively low. Mansouri et al. utilize this fact by employing an adaptive threshold against which to compare the power in the current epoch to determine whether it is a candidate for seizure. The adaptive threshold depends on the power in the prior epochs. This means that during non-seizure epochs with higher theta-alpha power, the adaptive threshold for declaring a seizure will also be higher, thus preventing false alarms.

If an epoch is declared to be a candidate for seizure, it is tested for hypersynchronocity. Treating the EEG leads as nodes in a network, two nodes are considered synchronized or connected if the Euclidean distance between the FFT coefficients in the high gamma band is below a certain threshold. The ratio of connected pairs to the total number of channel pairs provides a measure of hypersynchronization. If this measure exceeds a threshold, a seizure is declared. Historically, the clinical evaluation of EEG signals for seizures has not typically focused on higher frequencies. Part of the reason is historical. EEG waveforms were drawn on paper with a moving pen, and the pen could only move at a limited speed, so higher frequencies were not readily available to the clinician. Furthermore, the higher frequencies are often contaminated by electrical activity from skeletal muscles. However, when we take the Euclidean distance, it is reasonable to assume that the electrical activity due to the skeletal muscles will get canceled out, thus making this way of measuring hypersynchronization reasonable.

As one might expect, patient-specific algorithms generally perform better than patient-agnostic algorithms. Training the algorithm on the patient provides a much better differentiation between seizure and non-seizure epochs for that particular patient. So why even consider patient-agnostic algorithms? The reason is mainly resources. In many parts of the world, neurologists are in short supply, and their expertise is needed to provide initial seizure determinations, which can then be used to train the signal processing algorithm. This is especially true in low and middle-income countries, which account for about 80% of the people suffering from epilepsy [63]. Because of the lack of resources, the vast majority of the people suffering from epilepsy in these countries do not get identified and go untreated. According to the World Health Organization (WHO), almost 70% of these people would be symptom-free with anti-seizure medication, which is relatively cheap. Patient-agnostic algorithms such as the one described above can be implemented in simple and inexpensive processors. As dry electrodes for EEG become more widely available with a sharp drop in prices, a simple headband for seizure detection could be assembled at a low enough cost to make it widely accessible in low-resource environments.

8.4.3 Concussion

Traumatic brain injury, or TBI, is damage to the brain caused by a forceful impact. It is the fourth leading cause of brain injury, the first three being stroke, Alzheimer's disease, and Epilepsy. TBI can be rated on a number of scales, the most common being the Glasgow Coma Scale (GCS) [64] which scores individuals based on eye response (1 for inability to open eyes to 4 for spontaneously opening eyes), verbal response (1 for no response to 5 for normal conversation) and motor response (1 for no movements to 6 for obeying commands to move). A GCS score of 13 and above after an impact accounts for about 80% of all TBIs and is referred to as mild TBI (mTBI) or concussion. Concussion is thought to be more of a functional injury rather than a structural one, so it makes sense to try to identify functional changes in the brain in order to detect the occurrence of concussion. A very common way of non-invasively studying the functional response of the brain is through a method known as Event-Related Potential.

8.4.3.1 Event-Related Potential (ERP)

When we examine an EEG, much of the activity appears noise-like with a mean of zero. If we provide the brain with an input that requires a coherent response, while the response itself may not be noise-like, there is sufficient noise from all the other processes going on in the brain that it would be difficult to observe this coherent response in the presence of noise. If we repeatedly provide the brain with the same input, it is reasonable to assume that the response to that input will be the same each time. However, the noise-like activity in the background will be different each time. Therefore, if we average the EEG signals for these repeated inputs, there is a good chance that the noise-like activity will get averaged out. Let $e_k(t)$ be the observed EEG signal after the kth input. The EEG signal $e_k(t)$ will be the sum of the coherent response (the response to the input) $r(t)$ and the noise-like background activity $n_k(t)$

$$e_k(t) = r(t) + n_k(t)$$

If we now repeat this process for N times and take the average response $\bar{e}(t)$ we obtain

$$\begin{aligned}
\bar{e}(t) &= \frac{1}{N}\sum_{k=1}^{N}(r(t) + n_k(t)) \\
&= \frac{1}{N}\sum_{k=1}^{N} r(t) + \frac{1}{N}\sum_{k=1}^{N} n_k(t) \\
&= r(t) + \frac{1}{N}\sum_{k=1}^{N} n_k(t)
\end{aligned}$$

If we assume that the background EEG signal is truly noise-like with a zero mean, then the second term will go to zero, and we will obtain the response $r(t)$. Generally, we don't look for the entire response. Instead, we look at the first few negative or positive peaks. The names of the peaks are usually reported as N or P followed by either the time between stimulus and response or the order. Therefore, if the first negative peak occurs after 200 ms, it is referred to as N200 or N1.

Given that a concussion affects the brain's functioning, a reasonable way to measure the change in functioning would be to have a subject perform a standard cognitive task and then assess whether the resulting ERP, or perhaps specific components of the ERP, differ from expected values in a non-concussion situation. This has been the standard approach to detecting concussions.

For the ERP to be useful, we need a well-specified cognitive task. Several tasks have been used to elicit the ERP. One is the *oddball* task. In the oddball task, there are two types of stimuli: common and novel. For example, a visual oddball task may consist of presenting two different images to the subject: a box and a sphere. The subject is told to press a button when they see the sphere. Every T seconds, one of the two images is presented to the subject. Most of the time, it is the box, and sometimes, it is the sphere. What has been found is that the presentation of the novel image, in this case, the sphere, results in a substantial increase in the P300 signal in the frontal and parietal regions. In an auditory oddball task, we might have two different notes, one common and one novel, and the subject's task would be to indicate the presence of the novel input. Another commonly used task is the working memory task, known as the *n-back* task. In this case, the subject is asked to indicate whether the current stimulus is the same as a stimulus presented n trials earlier. A visual 2-back task might entail having participants view a screen on which uppercase letters of the English alphabet are presented at regular intervals. The participant's job is to indicate whether the current letter matches the letter presented two trials ago by pressing a button. As the participant performs this task multiple times, we can obtain multiple responses that we can then average to obtain the ERP. You can try out a visual 2-back task at the website https://www.psytoolkit.org/experiment-library/.

The change in ERP can be used to detect the occurrence of a concussion [65]. Mansouri et al. [66] use a measure of the similarity between the frequency components of EEG channels from a 2-back task as features. Features from the same individual are collected before and after an event that may have resulted in a concussion. All these features are combined into a feature vector, and a support vector machine is used to determine whether a concussion has occurred. They demonstrate that their method achieves about 97% accuracy. In general terms, this is the same approach that we repeatedly see. Use time-frequency analysis to generate features and then use a well-known classifier to make a decision. The novelty is always in the specifics of the problem. In this case, it is the recognition that the similarity measured between the signals in different EEG channels will be informative when the individual performs a defined task.

8.4.4 Sleep

Given the complexity of the brain and its degree of involvement in every aspect of our lives, it is understandable that there is a period during which the brain undergoes a process of cleanup and consolidation—namely, sleep. During sleep, there is a need for long-term disconnection of the sensory modalities from the cortex so that the brain can take care of all the things it needs to do to incorporate new information, perform maintenance of the system, and, in general, take care of all the things it could not do during the hectic wake times. The effect of this is to reduce some of the randomness of the EEG signal, making it a useful monitoring system for the brain. Why not just switch off all the senses? After all, we cut off, or seriously reduce, the amount of visual information by shutting our eyes. The problem is that, unlike visual input, other sensors, such as those for tactile input, depend on many, many more widely distributed sensors than the two localized visual sensors. For these sensors to change their functioning based on the system's overall state, that is, the body, the sensors would have to do a lot more work than they currently do. Right now, they only pass on the information they detect without considering the system's overall state. If the sensors had to consider the overall state of the system, it would make the system exponentially more complex, given the number of existing sensors.

Sleep is divided into 30-s epochs, during which the brain is monitored using EEG leads in the frontal, central, and occipital regions. Eye movement is monitored using an electro-oculogram, and muscle tone is monitored using an electromyogram. The waveforms from these monitoring systems are scored according to the rules of the Academy of Sleep Medicine [67]. The two broad categories of sleep are rapid eye movement (REM) sleep and non-rapid eye movement (NREM) sleep. As the name implies, REM sleep is characterized by rapid eye movement and is generally, though not necessarily [68], the stage in which most dreaming occurs. NREM sleep is further divided into stages. The number of NREM sleep stages was previously four but is now generally accepted as three. The first stage, N1 or NREM1, is a transitional state in which the person drifts between sleep and wakefulness. The EEG shows mostly alpha waves. In the N2 or NREM2 stage, the person is fully asleep, and the frequency of the EEG signals decreases, while their amplitude increases. This stage is characterized by the presence of sleep spindles (more on this later). The N3 or NREM3 stage is characterized by a further decrease in the frequency of EEG signals, which are now dominated by delta waves.

As we mentioned earlier, the thalamus serves as a central waystation to the cortex, making it an obvious location where we could interrupt the passage of sensory information. The brain uses a system of oscillations to somewhat isolate the cortex from the senses. These include slow oscillations below 1 Hz, sleep spindles, and delta waves. The slow oscillation first discovered by Steriade et al. [69] is in some ways the road on which the other oscillations travel. Steriade describes it as the coalescing factor. This oscillation is generated within the neocortex. We know this because when the thalamus is removed, the oscillation persists. However, when the cortex is removed, the oscillation disappears [70]. The slow oscillation consists of a prolonged depolarization or "UP" or "ON" phase followed by a long-lasting hyperpolarizing "DOWN" or "OFF" phase. By hyperpolarizing, we mean that the resting potential difference, or voltage, across the cell membrane of the neuron drops by more than 10 mV [53]. The hyperpolarization of the thalamocortical neurons serves to block, or perhaps attenuate, sensory signals from reaching the cortex [71]. The membrane potential in this hyperpolarized state is almost flat, indicating a lack of synaptic activity.

In the UP state, we see the emergence of *sleep spindles*. Spindles in humans are sequences in the EEG signal that recur rhythmically at frequencies ranging from 12 Hz (low-frequency spindles) to 15 Hz (fast-frequency spindles) and occur every 2–5 s during the NREM2 stage. They appear to be a side effect of the TRN's effort to prevent thalamus relay neurons from transmitting sensory information to the cortex. TRN neurons are rich in intrinsically bursting neurons. Rhythmic spike bursts from these neurons induce inhibitory post-synaptic potentials (IPSPs) in target thalamocortical neurons. These IPSPs result in burst firing in the neurons in the thalamus, which then re-excite the TRN neurons, creating a "ping-pong" effect recruiting increasingly larger groups of neurons in the thalamus [72]. We know that the TRN neurons are the pacemakers of spindles because spindles disappear in the thalamocortical system when the TRN is disconnected, but they are preserved in the TRN when it is disconnected from the thalamus and the cortex. What we observe in EEG recordings are excitatory post-synaptic potentials (EPSPs). What is important from the point of view of sleep are the IPSPs. These are what progressively hyperpolarize the thalamocortical neurons, thus disabling their relay functionality. Once the thalamocortical neurons are hyperpolarized, the feedback process ends, and the spindle disappears. Several other mechanisms have been proposed [72], perhaps acting in concert, for the termination of the sleep spindles. The thalamocortical neurons impinge on a broad range of cortical neurons. Therefore, it is reasonable that we would see spindles in various parts of the cortex. What is somewhat surprising is that these spindles are synchronized. Synchronization is performed not in the thalamus but in the cortex, indicating the strong feedback between cortical and thalamic neurons.

If you wish to play with some EEG signals, there are several sites on the internet from which you can download datasets. One of the best known is *physionet.org*. There is also public domain software for viewing and processing the EEG data. *FieldTrip* is an open source MATLAB toolbox [73], and you can find a tutorial for using this package to extract different sleep stages at [74].

8.4.5 Brain Computer Interface

A Brain-Computer Interface (BCI) translates brain activity into actions that a computer can take. This is particularly useful for individuals who suffer from neuromuscular disabilities [75]. This interaction using this interface varies from control of external devices to communication. While there are multiple ways to monitor brain activity, the noninvasive nature of the EEG means that it has always played a significant role in this area. The interfaces essentially require extracting features from brain signals and translating them into decisions. In a general sense, the paradigm remains the same one we have seen repeatedly: feature extraction followed by classification. An additional step of translating the classification into instructions may also be required. The features being classified can be time-domain features, such as evoked potential or firing rate, and they can also be frequency-domain features, such as the power in particular frequency regions. One of the first applications of BCI was for controlling a prosthetic hand [76]. Signals from the motor cortex obtained using scalp electrodes were shown to be capable of providing open/close signals to a prosthetic hand. Up until then, most attempts at controlling prosthetics had been made through the use of residual muscular or nerve activity in the vicinity of the amputated limb. This alternative approach, which utilized signals from the brain, required extensive processing, a task that was accomplished using a mainframe computer in those days. With the increased and increasing miniaturization of computational power, BCI applications for control have seen rapid growth [77, 78].

Jacques Vidal is credited with the term BCI, naming his laboratory at UCLA the "Brain Computer Interface Laboratory." His 1977 paper [79] describes a BCI system for moving a cursor through a maze. The subject is presented with the visual pattern similar to that shown in Fig. 8.19 along with a maze.

The stimulus consists of 30μs flashes illuminating the checkerboard pattern. The subject looks at one of the four circles, depending on the direction they want the cursor to move. Effectively, looking at the different circles puts the flash on a different part of the visual field. To collect the signal, the EEG electrodes are placed in locations Fz, Pz, Oz, $O1$, $O2$, and the

Fig. 8.19 Target presented to the subject

Inion. The electrode at Fz is used for artifact detection. A low-pass filter is used to smooth the voltages $Pz - Oz$, $O1 - Oz$, $I - Oz$, and $O2 - Oz$. The filtered signals sampled at 250 samples per second make up the initial feature vectors. An epoch is defined to be 0.4 s. This means that the initial feature vector has 400 components. As these components are highly correlated, a dimensionality reduction procedure (described in the paper) is used to reduce the feature vector to 10 dimensions. This feature vector is then classified into one of four classes corresponding to the directions up, down, right, and left. The classifier output controls the cursor, permitting the user to traverse the maze. This early application of BCI demonstrated the primary features that comprise most BCI applications. We begin with filtering and artifact removal. Then there is feature extraction, and finally, classification.

This particular application of BCI falls into the realm of *dependent* or *exogenous* BCI. The term dependent refers to the fact that the brain signal being used depends on some kind of muscular movement—in this case, the movement of the eyes. McFarland et al. [80] developed an independent, or endogenous, cursor control system by training the user to modulate EEG waveforms in the mu band (8–12 Hz) and the beta band (20–24 Hz) recorded over the sensorimotor cortex to move a cursor in the desired direction.

If a user is asked to look for a stimulus that occurs rarely, its appearance will elicit a wave 300–400 ms after its onset. This wave is known as the P300 wave and is utilized in several Brain-Computer Interface (BCI) algorithms. Farwell and Donchin [81] used this fact to develop a spelling algorithm that could be used by people with severe neuromuscular deficits. The 26 letters of the alphabet and a few commands were organized in a 6×6 table as follows. The user was told to focus on the letter (or command) they wanted, and the rows

A	G	M	S	Y	*
B	H	N	T	Z	*
C	I	O	U	*	TALK
D	J	P	V	FLN	SPAC
E	K	Q	W	*	BKSP
F	L	R	X	SPL	QUIT

and columns of the table were alternately flashed. When a row or column contained the letter the user was looking for, a P300 signal was elicited. By averaging over multiple trials, the algorithm determined which letter was being sought and printed the corresponding letter. If the algorithm's selection was inaccurate, the user could select BKSP (backspace) as their next choice. Once the complete word had been spelled out, the user could select TALK to activate a speech synthesizer. Farwell and Donchin [81] found that this system could operate at a rate of 12 bits/minute with an accuracy of 95%.

The paradigm introduced by Farwell and Donchin [81] has been extended and refined in many ways. Instead of doing a group-based selection, where the group is a row or a column, we can use a matrix in which individual letters are highlighted [82]. This results in an increase in the P300 peak when the correct letter is selected, thus increasing the accuracy.

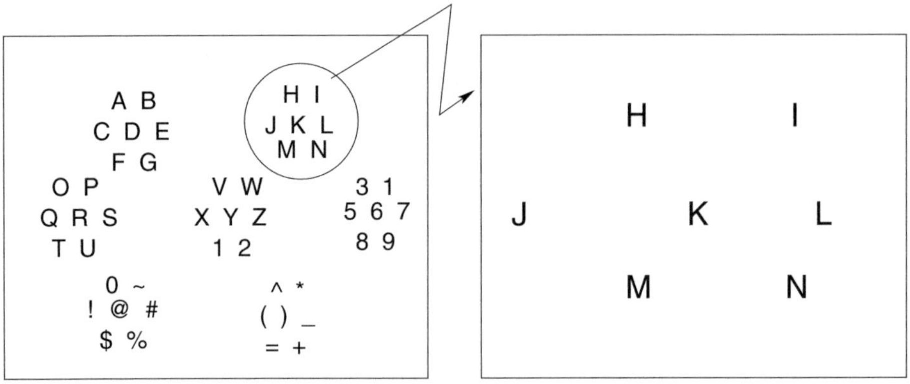

Fig. 8.20 A hierarchical system of selections [83]

Instead of the tabular arrangement, the target letters and words, or numbers, can be arranged in a hierarchical manner where a group is selected, followed by the selection of an individual element in a group. An example of this from [83] is shown in Fig. 8.20. If the second group in the top row is selected as the first selection, the user then selects from the elements of that group in the second selection. This method also decreases error rates. A recent extension of this approach is the Easy Screen P300 Speller of Aygün and Kavsaoğlu [84]. This application also uses two screens; the first screen is a 7×7 stimulus matrix. After the user has selected some letters, a list of twenty words beginning with those letters is presented to the user. The user can then move over to trying to select one of those words or continue the spelling process. This substantially speeds up the spelling process.

The use of a matrix or group of targets becomes problematic when the subjects struggle to shift their focus to different parts of the matrix. This is true of many late-stage ALS patients as well as others suffering from Locked-in syndrome (LIS). For these individuals, a variation that uses rapid serial visual presentation (RSVP) is more useful. In the RSVP paradigm, the letters are presented sequentially in the same location on the display. Acqualagna and Blankertz [85] present a scheme that uses the RSVP paradigm. They add some modifications to the basic approach. Noticing that certain letters, such as O and Q or K and X, can be easily mistaken for each other, they used a capital O and X and lowercase q and k. The other lowercase letters they used were c, d, f, i, p, v, w, y, z. They also added another feature, the N200 component. This is a negative peak that occurs 200–350 ms after the stimulus. They achieved accuracies of around 94 % at a rate of 1.43 symbols/minute. The performance of this system can be improved if it is used in conjunction with a statistical language model [86].

In our review, we have barely scratched the surface of the variety of BCI applications currently being explored. The area of BCI applications and techniques using EEG is expanding at an explosive rate, and signal processing is both an important assistant and driver of this explosion.

8.5 Summary

In this chapter, we examined one of the most widely studied physiological signals, the electroencephalogram. We began with a brief review of the organ that generates the signal, the brain, and then we examined how the signal is measured. We then looked at some of the basic processing steps. Finally, we examined several application areas for the EEG.

9 Computed Tomography

9.1 Introduction

Since the first X-ray of Anna Bertha Ludwig's hand in 1895, taken by her husband, Wilhelm Conrad Röntgen, the X-ray has been an invaluable tool in medicine, allowing us to look inside a patient. However, the standard X-ray provides only a two-dimensional view of the body. We would like to have a three-dimensional view, which is the goal of computed tomography. The word *tomography* comes from the Greek word *tomos* or slice. Tomography, then, is a method for creating a slice or cross-section of a three-dimensional object. The cross-sections can be assembled to get the three-dimensional representation. Computed tomography uses X-rays to provide the illumination and computational algorithms to combine the images generated by the X-rays into a three-dimensional image. The three-dimensional object, in our case, the human body, is irradiated by X-rays from multiple directions. Sensors around the body detect the degree of absorption of these X-rays. These absorption profiles are combined using a computational algorithm to construct a three-dimensional view of the body.

It could be argued [87] that the CT scanner is another gift to us from the Beatles. EMI, the Beatles' record label, made an enormous amount of money from their success. It invested some of that money into the work of the British electrical engineer Godfrey Hounsfield, who drew on his background in researching instruments of human destruction to create this instrument of human health in 1975. He was awarded the 1979 Nobel Prize in Physiology or Medicine for his invention.

The generation of absorption profiles can be understood using a theorem proposed by the Austrian (or Czech, depending on how you assign nationality) mathematician Johann Radon in 1917, which we examine in the next section. These profiles can be combined to develop an image slice using a method called backpropagation, which is discussed in the following section. The remainder of the chapter is devoted to signal processing techniques used to improve this image.

K. Sayood, *Biomedical Signal Processing*, Synthesis Lectures on Biomedical Engineering,
https://doi.org/10.1007/978-3-032-08855-0_9

9.2 The Radon Transform

Consider a three-dimensional object illuminated with X-ray beams. On the screen on the other side, each point results from an X-ray beam attenuated by the various tissues it has encountered in its journey through the body. Given many such points, we would like to reconstruct the tissues encountered in the X-ray's journey through the body. To see how the process works, consider an irregularly shaped two-dimensional object shown in Fig. 9.1. We position it with respect to a two-dimensional coordinate system.

Assume this object absorbs X-rays according to a function $f(x, y)$; the amount of X-rays absorbed by the location (x, y) is given by $f(x, y)$. We will irradiate this object from multiple directions to obtain an image of the object. First, let's consider the irradiation of the object from a single direction. Let's suppose we irradiate this object with X-rays in the y direction as shown in Fig. 9.2. The radiation is detected using sensors perpendicular to the direction of the rays. The amount of radiation the object absorbs can be obtained by integrating the absorption function $f(x, y)$ along the X-rays. If we denote the absorption profile by $p(x)$, then, for the ray that crosses the x axis at x_o, the absorption profile is given by

$$p_0(x_o) = \int_{-\infty}^{\infty} f(x_o, y)dy$$

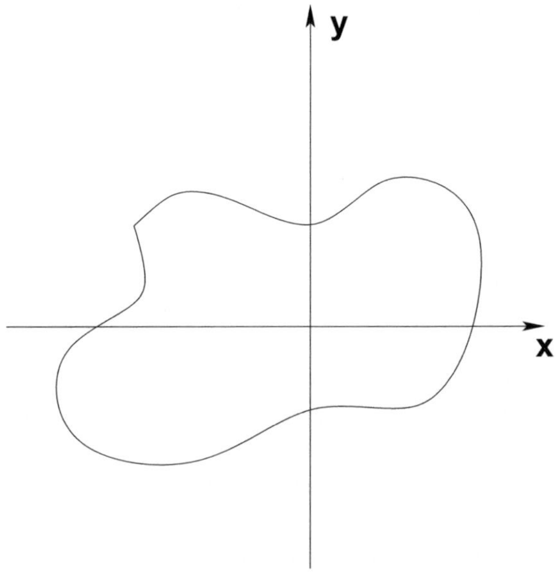

Fig. 9.1 An irregularly shaped object

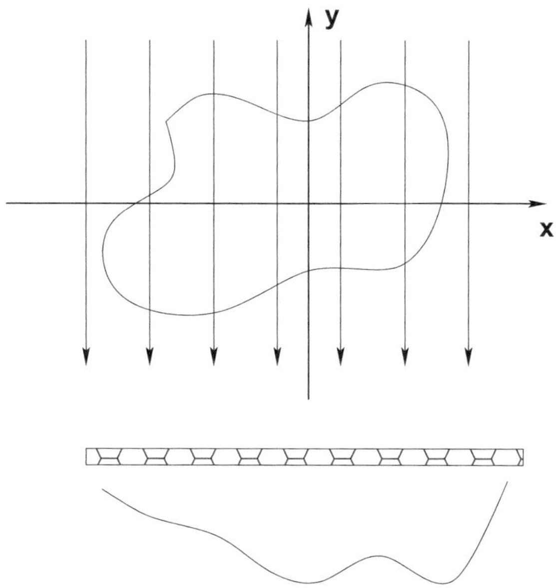

Fig. 9.2 An object irradiated by X-rays in the y direction

where the subscript indicates the angle of the sensor array with respect to the x axis. For a general value of x, we have

$$p_0(x) = \int_{-\infty}^{\infty} f(x, y)dy$$

This absorption profile provides only a minimal amount of information about the object's shape on its own. However, we are not restricted to irradiating the object in just one direction; we can irradiate it from multiple directions and obtain multiple absorption profiles.

It was easy to mathematically derive the absorption profile when the X-rays came from the y direction. We could just as easily derive the absorption profile if the X-rays were coming from the x direction; we would integrate $f(x, y)$ with respect to x instead of integrating with respect to y. But what if the illumination was not coming from either the x or the y direction as shown in Fig. 9.3?

To determine the absorption profile, we must integrate along the direction of the X-rays. In other words, we need to integrate $f(x, y)$ over all points (x, y) that fall on a particular beam. To see how to do that, let's define the new set of axes (shown in red), which are parallel and perpendicular to the rays. As we can see from Fig. 9.4, the new axes are at an angle of θ to the old axes. If the beam whose absorption profile value we are calculating is at a distance s from the origin, we can denote the absorption profile by $p_\theta(s)$.

Let's isolate one of the beams as shown in Fig. 9.5. To integrate $f(x, y)$ along this beam, we need to find an expression for the points (x, y) on this ray. We have marked a point (x, y) on the X-ray as a red dot. From the figure, we can see that the point with coordinates

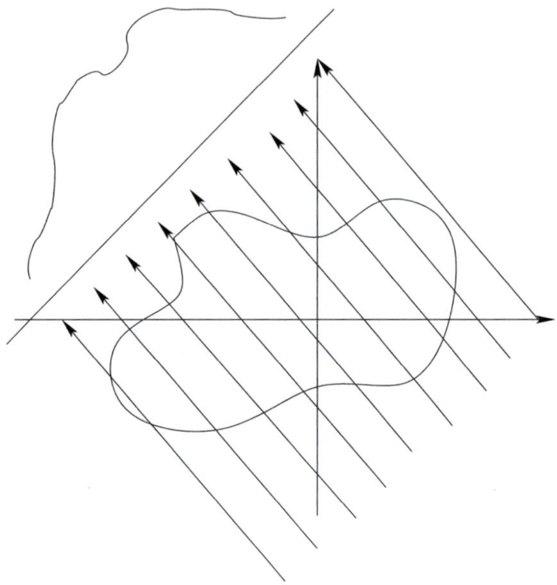

Fig. 9.3 The object irradiated from an arbitrary direction

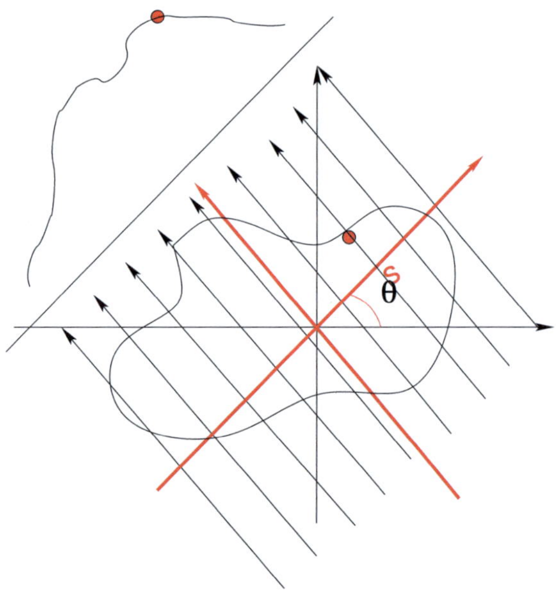

Fig. 9.4 An irregularly shaped object

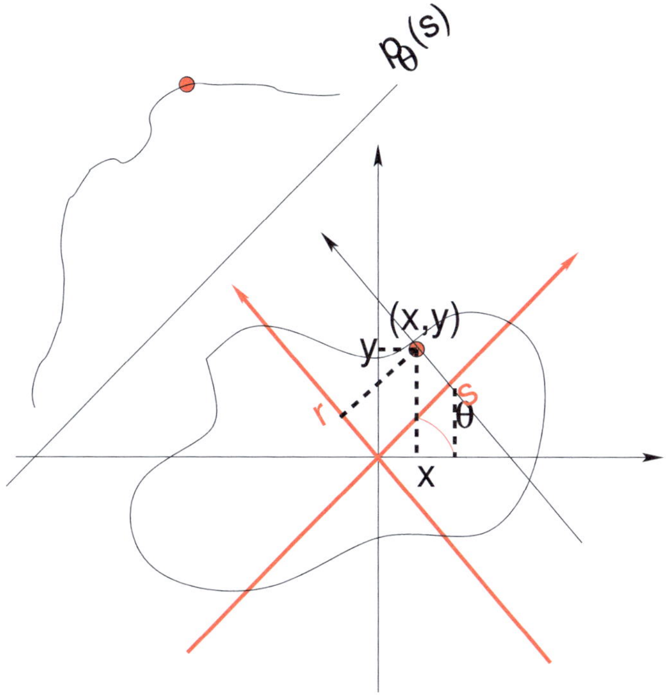

Fig. 9.5 The object irradiated from an arbitrary direction

(x, y) with respect to the original axes corresponds to the point (r, s) in terms of the new coordinate axes. Let's find r and s in terms of x and y. To do that, we first find x and y in terms of r and s.

$$x = s\cos(\theta) - r\sin(\theta)$$
$$y = s\sin(\theta) + r\cos(\theta)$$

Multiplying the top equation by $\cos(\theta)$ and the bottom equation by $\sin(\theta)$ we get:

$$x\cos(\theta) = s\cos^2(\theta) - r\sin(\theta)\cos(\theta)$$
$$y\sin(\theta) = s\sin^2(\theta) + r\cos(\theta)\sin(\theta)$$

Adding the two equations, we get

$$x\cos(\theta) + y\sin(\theta) = s(\cos^2(\theta) + sin^2(\theta)) = s$$

This means that all points (x, y) on the X-ray at a distance s have to satisfy the equation

$$x\cos(\theta) + y\sin(\theta) = s$$

where θ is the angle between the perpendicular to the beam and the x-axis. To get a point on the absorption profile, we need to integrate $f(x, y)$ over all points where $s = x\cos(\theta) + y\sin(\theta)$. We can do that by using the delta function.

If $f(x, y)$ is the absorption of the object at location (x, y), then the point on the projected absorption profile corresponding to the beam with parameters s and θ will be

$$p_\theta(s) = \int_{-\infty}^{\infty}\int_{-\infty}^{\infty} f(x, y)\delta(x\cos(\theta) + y\sin(\theta) - s)dxdy$$

This is called the *Radon transform*.

While we have discussed the mathematics of generating the absorption profile in the continuous domain, it is easier to see how things work in the discrete world, where integration becomes summation. So, let's use a toy example to see how this works in the discrete world. Let's suppose the absorption function for our object looks as follows:

1	9	5
6	0	1
9	5	7

The Radon transform for this at an angle of 0 degrees would be the summation of the absorption coefficients from top to bottom for each column.

16	14	13

The Radon transform for an angle of 90 degrees would be the summation of the absorption coefficients along each row.

15
7
21

The question now is, how do we recover the object from the absorption profile? The following section will address this question.

9.3 Back-Projection

To recover our original object, we will use a process called back-projection. Consider a situation where we have a discrete number of projections indexed by k. For a given angle θ_k the back-projection b_{θ_k} would be given by

$$b_{\theta_k}(x, y) = p_{\theta_k}(x\cos(\theta_k) + y\sin(\theta_k))$$

Summing up these back-projections over k would give us a representation of the original absorption coefficients

$$\hat{f}(x, y) = \sum_k b_{\theta_k}(x, y)$$

How good is this representation? Let's use our toy example. We start with a blank slate of the size of our original object.

Back-projecting the absorption profile at an angle of 0 means copying the 0-degree absorption profile to each row. As the absorption profile is the sum of the absorption of each pixel, we equally distribute the absorption value among the pixels. Therefore, the back-projected image from the single projection at an angle of 0 would be Similarly, the

16/3	14/3	13/3
16/3	14/3	13/3
16/3	14/3	13/3

back-projected image from the projection at 90 degrees would be where we copied the

15/3	15/3	15/3
7/3	7/3	7/3
21/3	21/3	21/3

column representing the absorption profile at an angle of 90 degrees into each column of the target image. Adding these, we get That doesn't look very much like our original object.

10.3	9.7	9.3
7.7	7	6.7
12.3	11.7	11.3

But then, we only had two projections to work with. Let's see what happens if we have many more projections to work with. It is easier to work with more projections if we use MATLAB, which is what we will do. First, let's generate an image to play with.

```
clear all
fxy = zeros(128);
for(i=1:1:128)
    for (j=1:1:128)
        k = (i-64)^2 + (j-64)^2;
        if k < 256
            fxy(i,j) = 255;
        end
```

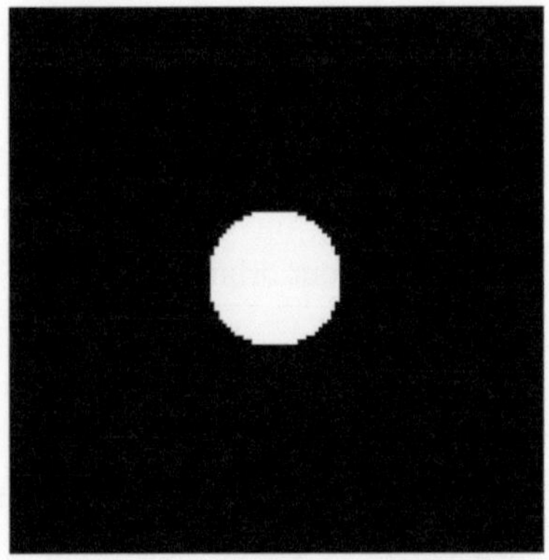

Fig. 9.6 A disc-shaped object

```
    end
end
figure
imshow(fxy,[])
```

This results in a white disk on a black background as shown in Fig. 9.6.

We are going to use the MATLAB functions *radon* and *iradon* to perform the Radon transform and back-projections. The function *radon* implements the Radon transform, that is, it generates the absorption profiles, and the function *iradon*performs the inverse, that is, it generates the back-projections. Let's take the radon transform for four values of θ: 0°, 45°, 90°, and 135°.

```
theta=0:45:179;
[p.s] = radon(fxy,theta);
```

We can display the radon transform as shown in Fig. 9.7, where we have exaggerated the θ axis. As one might expect, the view from the different angles is the same.

We can obtain a reconstruction of the original object using the *iradon* function

```
b = iradon(p,theta,'None');
figure
imshow(b,[])
```

With four back-projections, as we can see from Fig. 9.8, the resultant reconstruction, although not particularly good, is still recognizable. Let's increase the number of projections to eighteen by generating a projection every 10° as shown in Fig. 9.9.

This is much better. Though we still have blurriness around the edges.

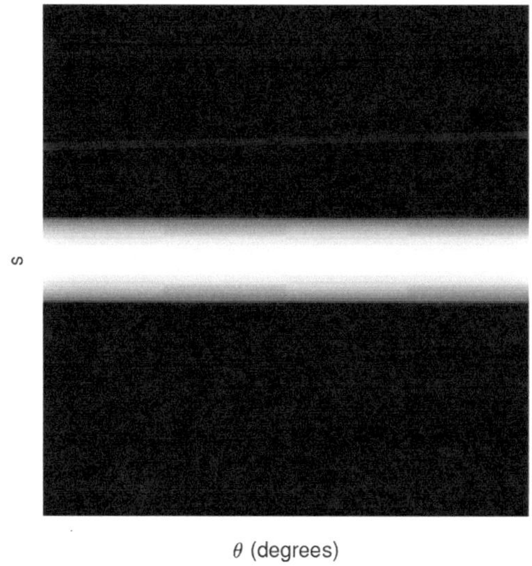

Fig. 9.7 The Radon transform of a disc-shaped object

Fig. 9.8 Reconstructed image of a disc using four projections

Instead of a circle, let's use a slightly more realistic image. Figure 9.10 shows a phantom image created by Larry Shepp and Benjamin Logan [88] for testing image reconstruction algorithms. In MATLAB, you can obtain the image by typing

```
fxy = phantom('Modified Shepp-Logan',200);
```

where 200 is the number of rows and columns of the image. The image is shown in Fig. 9.10

Using the same number of projections as before, the reconstructions are shown in Fig. 9.11. Clearly, four projections are nowhere close to enough for this slightly more com-

Fig. 9.9 Reconstructed image of a disc using eighteen projections

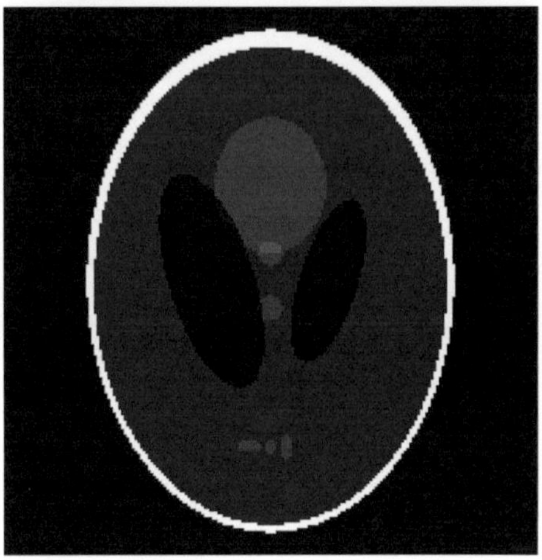

Fig. 9.10 The Shepp-Logan phantom

plex image. When we use eighteen projections, the result is significantly better, but there is still significant blurriness. To understand why this is the case, we need to examine the process in the frequency domain. We will do that with a theorem known as the *Fourier slice theorem*, or sometimes the *projection slice theorem* or the *central slice theorem*, which was initially developed to study a celestial source of radio waves [89].

9.4 The Fourier Slice Theorem

We begin with the Fourier transform of the projected absorption profile. Keeping the value of θ fixed, the one-dimensional Fourier transform of the projected absorption profile $p_\theta(s)$ is given by

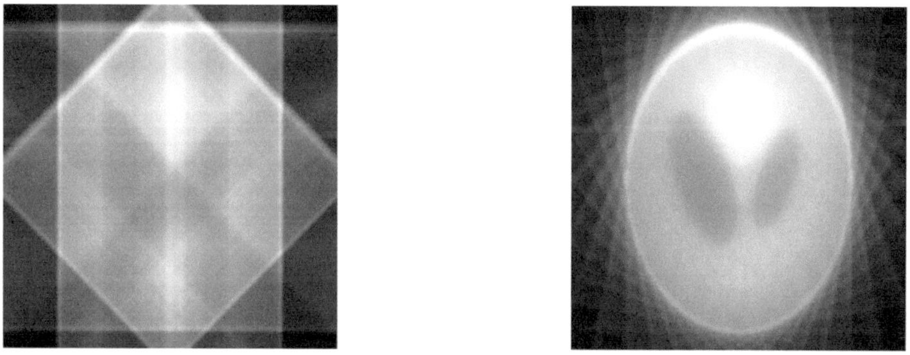

Fig. 9.11 Reconstruction of the Shepp-Logan phantom using four and eighteen projections

$$P(\omega, \theta) = \mathcal{F}\left[p_\theta(s)\right] = \int_{-\infty}^{\infty} p_\theta(s) e^{-j\omega s} ds$$

but from the Radon transform we saw earlier that

$$p_\theta(s) = \int_{-\infty}^{\infty}\int_{-\infty}^{\infty} f(x, y)\delta(x\cos(\theta) + y\sin(\theta) - s)dxdy$$

Substituting for $p_\theta(s)$ in our Fourier transform expression we get

$$P(\omega, \theta) = \int_{-\infty}^{\infty}\int_{-\infty}^{\infty}\int_{-\infty}^{\infty} f(x, y)\delta(x\cos(\theta) + y\sin(\theta) - s)dxdye^{-j\omega s}ds$$

Assuming all integrals converge, we can reverse the order of integration to get

$$P(\omega, \theta) = \int_{-\infty}^{\infty}\int_{-\infty}^{\infty} f(x, y)\left[\int_{-\infty}^{\infty} \delta(x\cos(\theta) + y\sin(\theta) - s)e^{-j\omega s}ds\right]dxdy$$

Using the sifting property of the delta function, this is

$$\begin{aligned} P(\omega, \theta) &= \int_{-\infty}^{\infty}\int_{-\infty}^{\infty} f(x, y)e^{-j\omega(x\cos(\theta)+y\sin(\theta))}dxdy \\ &= \int_{-\infty}^{\infty}\int_{-\infty}^{\infty} f(x, y)e^{-j(x\omega\cos(\theta)+y\omega\sin(\theta))}dxdy \end{aligned}$$

The two-dimensional Fourier transform of $f(x, y)$ is given by

$$F(u, v) = \int_{-\infty}^{\infty}\int_{-\infty}^{\infty} f(x, y)e^{-j(ux+vy)}$$

Looking at the expression for $P(\omega, \theta)$ we can see that

$$P(\omega, \theta) = F(u, v)\Big|_{u=\omega\cos(\theta) v=\omega\sin(\theta)}$$

That is, $P(\omega, \theta)$, the Fourier transform of the projection, is a slice through the two-dimensional Fourier transform of the original absorption profile. The sum of the back-projections then is the sum of the slices through the two-dimensional Fourier transform. If we picture this summation of slices as shown in Fig. 9.12, we can see the reason for the blurring and the halo around the sphere in our reconstruction. We can see from Fig. 9.12 that there is an overrepresentation of low frequencies and an undersampling of high frequencies. The latter leads to a reduction in the definition of the edges, while the former results in a smearing-out effect.

Now that we know where the problem lies, we can see how to fix it. The fix, for obvious reasons, is called *filtered back-projection*.

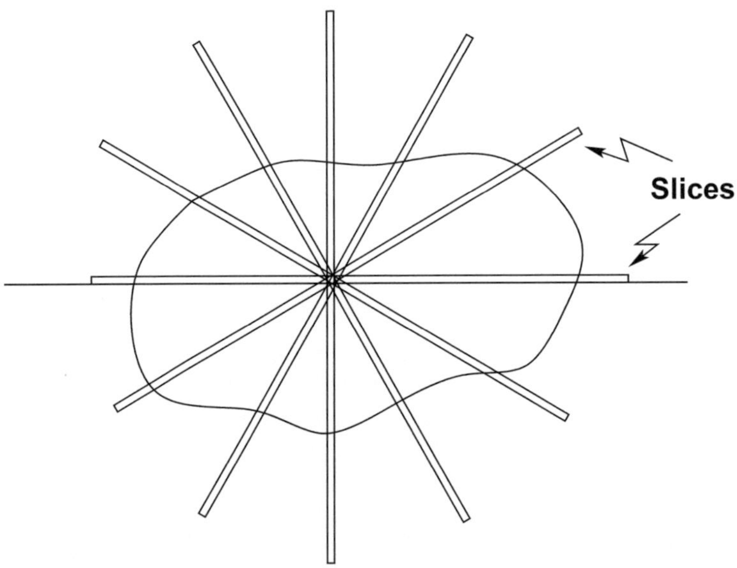

Fig. 9.12 Slices of the Fourier transform

9.5 Filtered Back-Projection in the Fourier Domain

Let us begin with the inverse Fourier transform of $F(u, v)$

$$f(x, y) = \frac{1}{(2\pi)^2} \int_{-\infty}^{\infty} \int_{-\infty}^{\infty} F(u, v) e^{j(ux+vy)} du dv$$

Let's do a variable substitution

$$u = \omega \cos(\theta)$$
$$v = \omega \sin(\theta)$$

This looks like going from Cartesian to polar coordinates (think r instead of ω), and as was the case there

$$du dv = \omega d\omega d\theta$$

with the limits changed accordingly. Using these substitutions, we can write the equation for $f(x, y)$ as

$$f(x, y) = \frac{1}{(2\pi)^2} \int_{0}^{2\pi} \int_{-\infty}^{\infty} F(\omega \cos(\theta), \omega \sin(\theta)) \omega d\omega d\theta$$

but

$$F(\omega \cos(\theta), \omega \sin(\theta)) = P(\omega, \theta)$$

Therefore,

$$f(x, y) = \frac{1}{(2\pi)^2} \int_{0}^{2\pi} \int_{-\infty}^{\infty} \omega P(\omega, \theta) d\omega d\theta$$

Instead of integrating from 0 to 2π we can change the limits of integration to

$$f(x, y) = \frac{1}{(2\pi)^2} \int_{0}^{\pi} \int_{-\infty}^{\infty} |\omega| P(\omega, \theta) d\omega d\theta$$

This is the inverse Fourier transform of the product of $|\omega|$ and $P(\omega, \theta)$, which is the same as the convolution of the inverse Fourier transform of $|\omega|$ and $P(\omega, \theta)$. The inverse Fourier transform of $P(\omega, \theta)$ is $p_\theta(s)$, which we already possess. Obtaining the inverse Fourier transform of $|\omega|$ is another matter. The function $|\omega|$ is not absolutely integrable and hence does not have an inverse Fourier transform. Therefore, we need to look for workarounds. A straightforward method is to truncate the ramp function; this is implemented as the option *Ram-Lak* in MATLAB. However, this will still result in the amplification of any high-frequency noise that exists in the projection. In practice, this is almost always the case. To avoid this, we can multiply the ramp function by a window. A common window to use is the

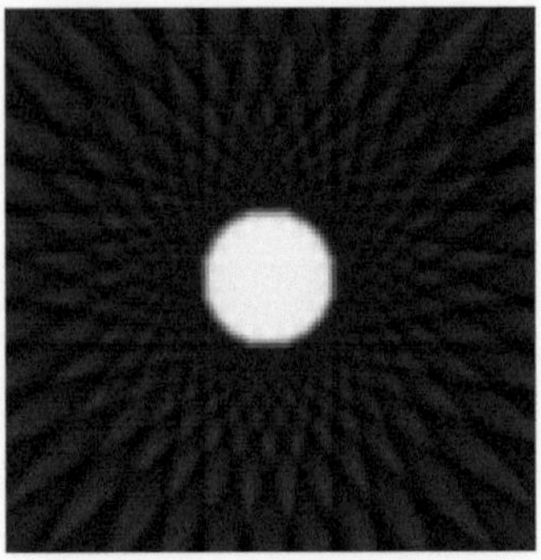

Fig. 9.13 Comparing filtered back projection using the truncated ramp filter (left) and the ramp multiplied with the Hamming window (right)

Fig. 9.14 Reconstruction using the ramp filter multiplied by the Hamming window

Hamming window. To observe the effect of high-frequency amplification, consider the two reconstructions shown in Fig. 9.13, where we have used eighteen projections. In the image reconstructed using the truncated ramp window, we can see the Moire patterns within the circle resulting from the high-frequency distortion.

If we use 60 projections and then use the Hamming window multiplied ramp filter for the back projections, we get the result shown in Fig. 9.14.

The reconstruction, while not perfect, is quite good.

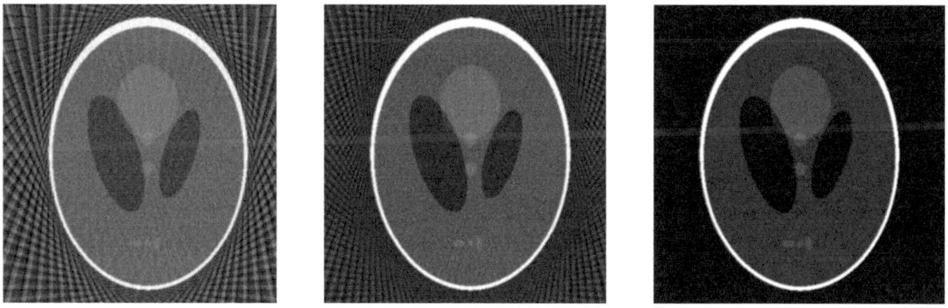

Fig. 9.15 Reconstructions of the Shepp-Logan phantom (from left to right) using 36 projections, using 80 projections, and 180 projections

9.5.1 How Many Projections Do We Need?

When we increased the number of projections from four to eighteen to 60, the quality of our reconstruction of the white disc increased dramatically. In fact, simply increasing the number of projections from four to eighteen resulted in a significant improvement in the quality of the reconstruction. Going from four to eighteen projections improved the reconstruction of the Shepp-Logan phantom as well, but the result with the eighteen projections was not as satisfactory. If we increase the number of projections further, the situation improves as shown in Fig. 9.15. The reconstruction from 36 projections clearly has some issues; the reconstruction using 80 projections is significantly improved, but you can still see the moire patterns around the boundary. With 180 projections, we get a reconstruction that is almost identical to the original without moire patterns.

What if we took a more realistic image, such as the one in Fig. 9.16? Using the same number of projections as we did for the Shepp-Logan phantom, we get the reconstructions shown in Fig. 9.17.

While the reconstruction with the 180 projections is significantly improved over the reconstructions using 36 projections and 80 projections, the quality of the reconstruction is still not close to the original. It takes about 3,600 projections to obtain a reconstruction visually identical to the origin. Why is there a difference between the required number of projections to get a visually close reconstruction? If we go back to the Fourier slice theorem, we can see that each projection is a sample of the Fourier domain. Recall that, according to the Nyquist sampling theorem, the number of samples we need in the time domain depends on the range of frequencies in the Fourier domain. Similarly, the number of Fourier slices, and hence the number of projections, depends on the spatial variations of the cross-section being imaged. So, the more detailed the cross-section, the more projections we will need. We can use the size of the cross-section in terms of pixels as a proxy for spatial variation - the more pixels there are, the more the likelihood of spatial variation.

Fig. 9.16 Two of the reviewers for this book in an earlier stage of their lives

Fig. 9.17 Reconstructions of the *reviewers* image (from left to right) using 36 projections, using 80 projections, and 180 projections

Here we run into a problem. For images, we can increase the number of projections with only increased computational costs. However, in real life, when the object being imaged is a human being, the increase in the number of projections results in a significant increase in the radiation dosage to which the patient is subjected. This can be dangerous. Fortunately, a computational technique is available that can provide improved resolution without necessarily increasing the risk of high doses of radiation. The method is called *iterative reconstruction*. The technique has been known for a while, but it became practical to use with the increased availability of computational power.

9.6 Iterative Reconstruction

The idea behind iterative reconstruction algorithms is simple. The only thing we know about the object is the projections. Therefore, our reconstruction should be such that if we use the reconstruction to form projections, instead of the original object, these projections should be the same as those obtained from the original object. So the process is:

1. Construct a back-projected or filtered back-projected image
2. Project through the constructed image to create artificial projections.
3. Use the difference between the actual and artificial projections to update the back-projected image.
4. Repeat until the difference between the actual and artificial projections is below an acceptable threshold.

To develop a computationally efficient algorithm, we require a mathematical framework that enables the efficient computational implementation of the projection operation. To do this, we take our $N \times N$ back-projected image X and write it as a $1 \times 2N$ column vector. The projection operation can then be implemented as a matrix multiplication. For example, consider the following 2×2 "image" f.

5	7
6	2

If we assume each pixel contributes equally to the projection, we can obtain the following six projections.

	5	7	***12***
	6	2	***8***
13	***11***	***9***	***7***

If we write our image and projections as column vectors

$$f = \begin{bmatrix} 5 \\ 7 \\ 6 \\ 2 \end{bmatrix}; \quad P = \begin{bmatrix} 12 \\ 8 \\ 7 \\ 9 \\ 11 \\ 13 \end{bmatrix}$$

we can define a projection matrix

$$A = \begin{bmatrix} 1\,1\,0\,0 \\ 0\,0\,1\,1 \\ 1\,0\,0\,1 \\ 0\,1\,0\,1 \\ 1\,0\,1\,0 \\ 0\,1\,1\,0 \end{bmatrix}$$

such that

$$P = Af$$

In this case, we knew f and we used that to find P. In reality, we do not know the true absorption profile; we usually have noise in the system, and we have assumed that the locations we have conceptualized as pixels all contribute equally to the projection. For now, let's disregard the last two and focus on the first. The only thing we know for sure is the projection values. The other thing we can obtain is an estimate of f; let's call that $f^{(0)}$. Given this estimate, we can try to find the actual value of f. We can pose this problem as that of finding f such that the difference between the true projection and the projection obtained from the estimate is minimized. In other words, find f such that $||Af - P||^2$ is minimized. When we want to minimize something, we can do that by taking the derivative and setting it equal to zero. In this case, we are dealing with vectors and matrices, so the derivative becomes the gradient. Expanding $||Af - P||^2$

$$\begin{aligned} ||Af - P||^2 &= (Af - P)^T(Af - P) \\ &= f^T A^T Af - f^T A^T P - P^T Af + P^T P \\ &= f^T A^T Af - 2f^T A^T P + P^T P \end{aligned}$$

and then, setting the gradient with respect to f to zero, we get

$$2A^T Af - 2A^T P = 0$$

from which we get

$$f = (A^T A)^{-1} A^T P$$

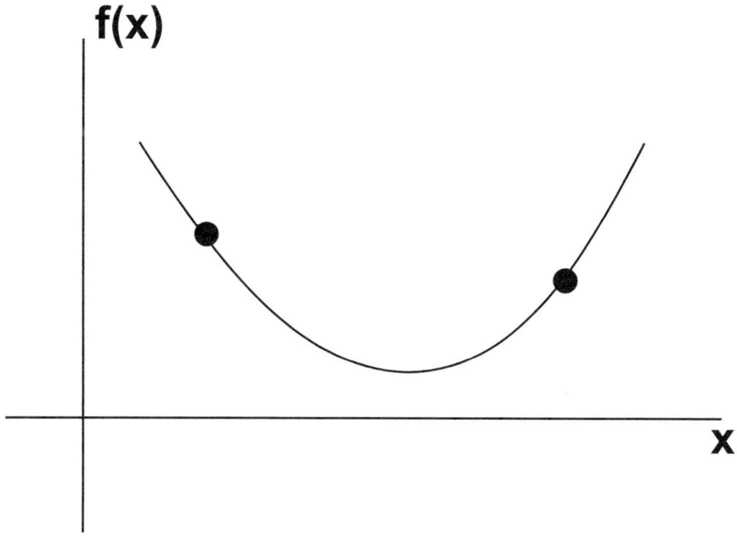

Fig. 9.18 A function with a single minimum

There are several problems with this solution. For a 512×512 image, the dimension of f is greater than half a million. Even if that was not a problem, the inverse of $A^T A$ generally does not exist. Therefore, instead of finding an exact solution, we need to find an approximate solution. A standard way to do this is to use a *gradient descent* algorithm.

To understand how the gradient descent algorithm works, consider the simple function shown in Fig. 9.18.

Let's suppose we guess the minimum of this function. In this simple case, the guess could be to the right of the actual minimum, or it could be a point to the left of the actual minimum. If our guess is to the left of the minimum, the derivative at this point will be negative, and to get to the minimum, we need to add a positive value to our guess. If our guess is to the right of the actual minimum, the derivative at this point will be positive, and we have to add a negative value to our guess. In other words, we will always have to add a value to our estimate with a sign that is proportional to the negative of the derivative.

In the general n-dimensional case, we replace the derivative with the gradient, but otherwise the approach remains the same. In the case of the problem we are looking at, this means that our reconstruction estimates will proceed as

$$f^{(k+1)} = f^{(k)} - \mu A^T (A f^{(k)} - P)$$

where f^k is the kth estimate of f. Notice that we are taking the difference between the projection generated by the reconstruction $f^{(k)}$ and the actual projection, and back-projecting it by premultiplying it with A^T to adjust the current reconstruction.

To see how this works in practice, let's use our simple 2×2 example. Let's pick $\mu = 0.1$ and let

$$f^{(0)} = \begin{bmatrix} 4 \\ 5 \\ 5 \\ 2 \end{bmatrix}$$

then

$$Af^{(0)} = \begin{bmatrix} 9 \\ 7 \\ 6 \\ 7 \\ 9 \\ 10 \end{bmatrix}; \quad Af^{(0)} - P = \begin{bmatrix} 9 \\ 7 \\ 6 \\ 7 \\ 9 \\ 10 \end{bmatrix} - \begin{bmatrix} 12 \\ 8 \\ 7 \\ 9 \\ 11 \\ 13 \end{bmatrix} = \begin{bmatrix} -3 \\ -1 \\ -1 \\ -2 \\ -2 \\ -3 \end{bmatrix}$$

Premultiplying this with A^T we get

$$\begin{bmatrix} 1\,0\,1\,0\,1\,0 \\ 1\,0\,0\,1\,0\,1 \\ 0\,1\,0\,0\,1\,1 \\ 0\,1\,1\,1\,0\,0 \end{bmatrix} \begin{bmatrix} -3 \\ -1 \\ -1 \\ -2 \\ -2 \\ -3 \end{bmatrix} = \begin{bmatrix} -6 \\ -8 \\ -6 \\ -4 \end{bmatrix}$$

Multiplying this with $\mu = 0.1$ and subtracting it from f^0 we get

$$f^{(1)} = \begin{bmatrix} 4.6 \\ 5.8 \\ 5.6 \\ 2.4 \end{bmatrix}$$

Round this to the closest integer, and we get an approximation for f

$$\hat{f} = \begin{bmatrix} 5 \\ 6 \\ 6 \\ 2 \end{bmatrix}$$

which is quite close to the actual value.

This is a simple form of iterative reconstruction. We can make it more complex and sophisticated in several ways. The adaptation step size μ influences the speed of adaptation; a larger value of μ means faster convergence, but it also results in less accurate location estimates. If we keep μ small, then the convergence is slower, but the system will converge to a point closer to the actual minimum. We can try to have the best of both worlds by making the step size a function of the iteration index k. We can start with μ_k large for smaller values of k and make it smaller as k increases.

In this formulation, we have made the error the Euclidean distance between the actual projection and the projection using the reconstruction. We can instead weight the errors by replacing $||Af - P||^2$ with

$$||Af - P||_W^2 = (Af - P)^T W(Af - P)$$

where W is a diagonal noise weighting matrix.

We could also add a regularization term to deal with undesired effects. If we keep the step size fixed and add the regularization term, we end with

$$f^{(k+1)} = f^{(k)} - \mu \left[A^T W(Af^{(k)} - P) - \nu Rf^{(k)} \right]$$

which is known as the Landweber method [90].

Procedures that step along an axis, evaluate some measure of error, and adjust accordingly are descended from a method for finding the roots of a polynomial developed by the Persian mathematician Muhammad Ibn Musa Al-Khwarizmi, who worked in the House of Wisdom in Baghdad during the Abbasid Caliphate. His name "Al-Khwarizmi," is the source of the word *algorithm*.

9.6.1 Algebraic Iterative Reconstruction (ART)

A different approach to the reconstruction is to set up the projection as a set of linear algebraic equations, which can then be solved iteratively. We can do this by writing the projection operation

$$Af = P$$

as a set of equations

$$\sum_j a_{i,j} f_j = p_i, \quad i = 1, 2, \ldots, M$$

To see how we can iteratively solve these equations, consider a set of two equations

$$a_{11} f_1 + a_{12} f_2 = p_1$$
$$a_{21} f_1 + a_{22} f_2 = p_2$$

Each of these equations is the equation of a line. The solution of these simultaneous equations is the point at which they intersect. An iterative method for solving these equations is to start with a guess, then project the guess onto one of these lines. The projection is then projected onto the other line and then back to the first one. The sequence of iterative projections will converge increasingly closer to the solution. The process is shown in Fig. 9.19.

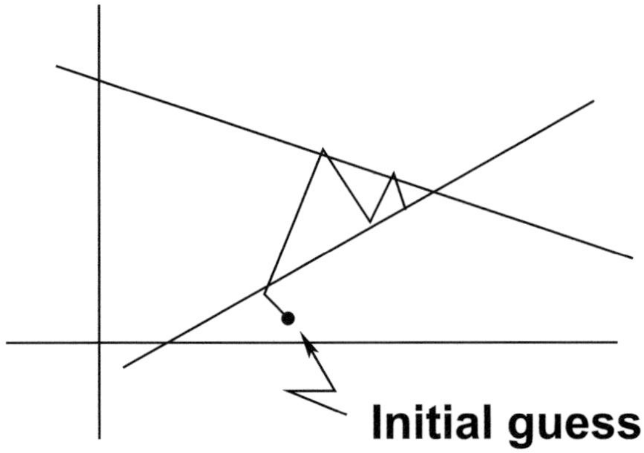

Fig. 9.19 Iterative solution of two simultaneous linear equations

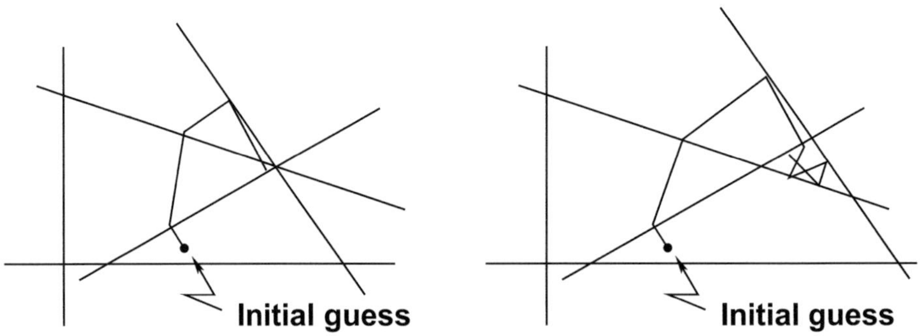

Fig. 9.20 Iterative solution of three simultaneous linear equations when they are consistent and when they are not

Now, let's add a third equation. With three equations, we could end up with a unique solution - the three lines intersect at a point. Alternatively, we might have a situation without a unique solution, where pairs of lines intersect at different points. Graphically, we can show these as shown in Fig. 9.20.

To see how we can implement this iterative projection algorithm, let's begin with a geometric interpretation of the equation for a line. Consider the line

$$y = 2x - 1$$

We can rewrite this as

$$2x - y = 1$$

or as a dot product of vectors

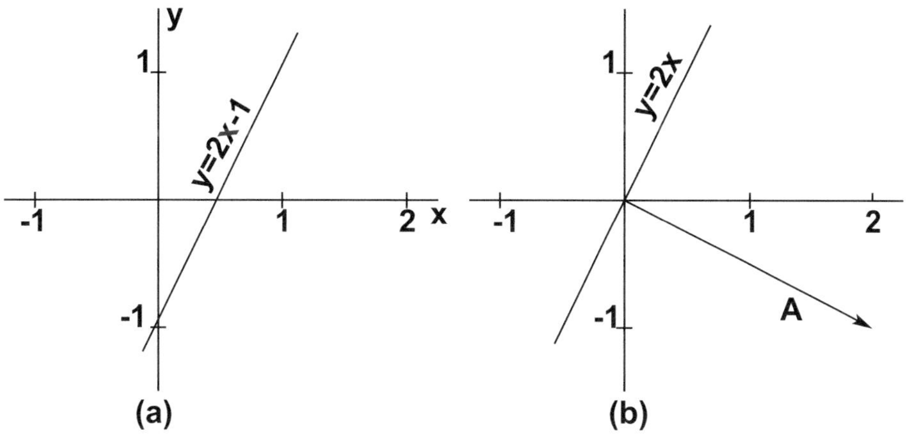

Fig. 9.21 The line of $y = 2x - 1$

$$\mathbf{A} \cdot \mathbf{x} = \begin{bmatrix} 2 \\ -1 \end{bmatrix} \cdot \begin{bmatrix} x \\ y \end{bmatrix} = 1$$

shown in Fig. 9.21a.

To see the relation between **A** and the line, we can set the intercept value to 0, or equivalently slide the line to the origin as in Fig. 9.21b. Setting the intercept value to 0 gives us the equation

$$\mathbf{A} \cdot \mathbf{x} = 0$$

from which we can see that the vector **A** is orthogonal to the line.

Let's now utilize this relationship to develop our algorithm for solving the equations that comprise the projection operation. We will do this for two dimensions and then generalize. Let's rewrite the equations

$$a_{11} f_1 + a_{12} f_2 = p_1$$
$$a_{21} f_1 + a_{22} f_2 = p_2$$

as

$$\mathbf{A}_1 \cdot \mathbf{f} = p_1$$
$$\mathbf{A}_2 \cdot \mathbf{f} = p_2$$

We begin with an initial guess, $f^{(0)}$, and then project this onto our first line to obtain the updated estimate, $f^{(1)}$. In Fig. 9.22 we have represented the estimates $f^{(0)}$ and $f^{(1)}$ as vectors, and we have drawn only one of the lines to simplify the image.

In Fig. 9.22a, the vector **c** is the vector from $f^{(1)}$ to $f^{(0)}$. Therefore,

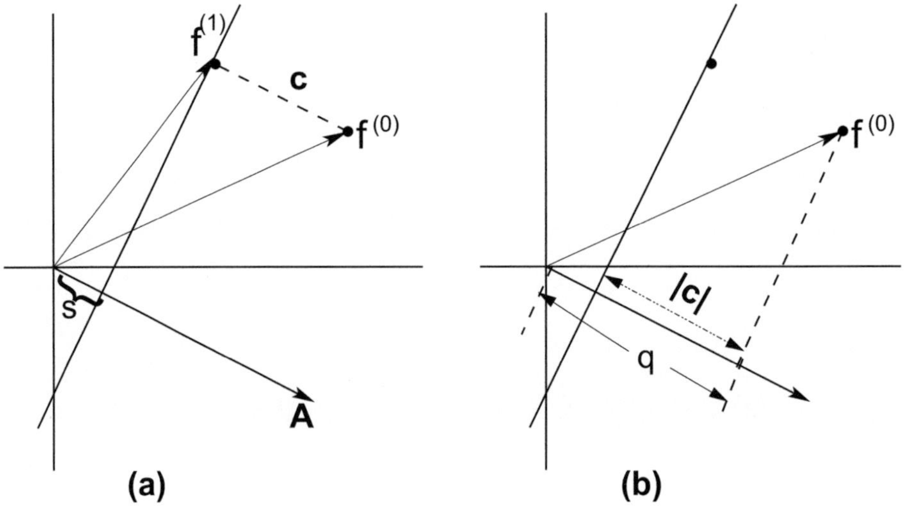

Fig. 9.22 Constructing a relationship between the updated estimate $f^{(1)}$ and the initial estimate $f^{(0)}$

$$f^{(1)} = f^{(0)} - \mathbf{c}$$

Recall that we obtain $f^{(1)}$ by dropping a perpendicular on the line from $f^{(0)}$, therefore, $\mathbf{c}$ is perpendicular to the line. But from our earlier development, we know that $\mathbf{A}_1$ is perpendicular to the line. Therefore, we can write $\mathbf{c}$ in terms of $\mathbf{A}_1$. Now we know the direction of $\mathbf{c}$, but what is the magnitude of $\mathbf{c}$?

We are going to express $\mathbf{c}$ in terms of the projections of $f^{(1)}$ and $f^{0)}$ on the vector $\mathbf{A}_1$. The projection of $f^{(1)}$ on $\mathbf{A}$, denoted by s, can be found by taking the dot product of $f^{(1)}$ with the unit vector in the $\mathbf{A}_1$ direction.

$$s = f^{(1)} \cdot \frac{\mathbf{A}_1}{\sqrt{\mathbf{A}_1 \cdot \mathbf{A}_1}}$$

In fact, s is the dot product of any point on the line with the unit vector in the $\mathbf{A}_1$ direction. Denoting a general point on the line by $\mathbf{f}$,

$$s = \mathbf{f} \cdot \frac{\mathbf{A}_1}{\sqrt{\mathbf{A}_1 \cdot \mathbf{A}_1}} = \frac{\mathbf{f} \cdot \mathbf{A}_1}{\sqrt{\mathbf{A}_1 \cdot \mathbf{A}_1}} = \frac{p_1}{\sqrt{\mathbf{A}_1 \cdot \mathbf{A}_1}}$$

Similarly the projection of $f^{(0)}$ on $\mathbf{A}_1$, denoted by q, can be found by taking the dot product of $f^{(0)}$ with the unit vector in the $\mathbf{A}_1$ direction.

$$q = f^{(0)} \cdot \frac{\mathbf{A}_1}{\sqrt{\mathbf{A}_1 \cdot \mathbf{A}_1}}$$

The magnitude of the vector $\mathbf{c}$ is simply the difference between q and s, while the direction of the vector $\mathbf{c}$ is $\mathbf{A}_1$.

$$\mathbf{c} = (q - s)\frac{\mathbf{A}_1}{\sqrt{\mathbf{A}_1 \cdot \mathbf{A}_1}}$$

Putting everything together, we have

$$\mathbf{c} = \frac{(f^{(0)} \cdot \mathbf{A}_1 - p_1)}{\sqrt{\mathbf{A}_1 \cdot \mathbf{A}_1}} \frac{\mathbf{A}_1}{\sqrt{\mathbf{A}_1 \cdot \mathbf{A}_1}} = \frac{(f^{(0)} \cdot \mathbf{A}_1 - p_1)}{\mathbf{A}_1 \cdot \mathbf{A}_1}\mathbf{A}_1$$

and

$$f^{(1)} = f^{(0)} - \frac{(f^{(0)} \cdot \mathbf{A}_1 - p_1)}{\mathbf{A}_1 \cdot \mathbf{A}_1}\mathbf{A}_1$$

Generalizing this to an arbitrary k

$$f^{(k+1)} = f^{(k)} - \frac{(f^{(k)} \cdot \mathbf{A}_{k+1} - p_{k+1})}{\mathbf{A}_{k+1} \cdot \mathbf{A}_{k+1}}\mathbf{A}_{k+1}$$

Once we reach the Mth projection, we can set the $f^{(0)} = f^{(M)}$, and restart the process. We can use the magnitude of the difference between consecutive projections as the stopping criterion.

Recall that $\mathbf{A}$ is the projection matrix describing the image generation process, and $\mathbf{f}$ is the vector of pixels. As the elements of the projection matrix $\mathbf{A}$ are ones and zeros, the computation process is straightforward. For example, $\mathbf{A}_k \cdot \mathbf{A}_k$ is simply the number of ones in kth row of $\mathbf{A}$, and to compute $f^{(k-1)} \cdot \mathbf{A}_k$ we simply add the components of $f^{(k-1)}$ which correspond to ones in the kth row of $\mathbf{A}$.

9.6.2 Commercial Iterative Systems

The increase in computational power, which made the use of iterative algorithms feasible, was a significant boost to the area of computed tomography. The ability to reduce the radiation dose to patients made the use of CT more attractive. This led to a significant proliferation of iterative reconstruction-based devices from a variety of manufacturers. Because the methods are proprietary, we do not have exact descriptions of the various algorithms. However, they are all based on the same principle - iteratively adjust the reconstruction so that the projections obtained from the scanning process and the projections that the reconstructions would have generated are similar. Here is a listing of some of the more popular systems.

1. Adaptive Statistical Iterative Reconstruction (ASIR) (GE)
2. Model Based Iterative Reconstruction (MBIR) (GE)
3. Iterative Reconstruction in Image Space (IRIS) (Siemens)
4. Sinogram Affirmed Iterative Reconstruction (SAFIRE) (Siemens)

5. Advanced Model Based Iterative Reconstruction (ADMIRE) (Siemens)
6. Iterative Model Reconstruction (IMR) (Philips)
7. Adaptive Iterative Dose Reduction (AIDR) Toshiba
8. SafeCT (Medic Vision)

9.7 Summary

In this chapter, we have examined how to reconstruct a slice through a three-dimensional object, such as the human body, using a sequence of two-dimensional X-ray images. We looked at signal processing issues related to the reconstruction process. Then we saw how, by posing the problem mathematically, we could arrive at a much safer, and therefore, more practical solution to the reconstruction problem.

Magnetic Resonance Imaging 10

We see something because photons bounce off a surface and enter our eyes, interacting with the rods and cones to generate the image. We perceive the object in different colors because different surfaces absorb different frequencies of light. In Magnetic Resonance Imaging (MRI), the imaging sensors perceive the nuclear magnetic resonance from certain types of atoms, specifically, those with an imbalance between neutrons and protons in their nucleus. This idea was behind nuclear magnetic resonance (NMR) spectroscopy. Still, it is one thing to detect the presence or absence of certain isotopes and another to map their positions within a living body. The living qualifier is crucial because it substantially reduces the strength of the magnetic field that can be used; NMR spectroscopy typically uses field strengths an order of magnitude greater than those used for MRI, which would cook the human flesh. MRI should be called (and was for a while) NMRI, but the word "nuclear" was dropped from the name of this modality because manufacturers thought patients might be squeamish about being put in a machine with the word "nuclear" in its name. To image living tissue at field strengths that are not detrimental to life, the MRI system combines quantum mechanics, electromagnetic theory, signal processing, and considerable ingenuity.

10.1 Magnetic Moments and Resonance

A spinning nuclear particle will generate a magnetic moment with a direction determined by the right-hand rule. The reason for the odd number requirement is that the Pauli Exclusion Principle dictates that nuclear particles on the same energy level will have opposite spins and opposite magnetic moments, which cancel each other out. Therefore, elements with an even number of protons and neutrons do not have a net spin. The hydrogen ion, being a single proton, generates a magnetic moment. This is quite serendipitous because the tissues

K. Sayood, *Biomedical Signal Processing*, Synthesis Lectures on Biomedical Engineering,
https://doi.org/10.1007/978-3-032-08855-0_10

we want to image are rich in water, which is the source of the hydrogen atoms whose nuclear resonance we are detecting.

The spinning nucleus generates a magnetic moment or a *nuclear magnetic dipole moment* $\vec{\mu}$ which is related to the spin angular momentum $\vec{J}$ by

$$\vec{\mu} = \gamma \vec{J}$$

where γ is known as the *gyromagnetic ratio*. The magnitude of the nuclear magnetic dipole moment is given by

$$|\vec{\mu}| = \gamma \bar{h}\sqrt{I(I+1)}; \quad \bar{h} = \frac{h}{2\pi}$$

where h is the Planck's constant, and I is the *spin quantum number*. The spin quantum number is a half integer ($\frac{1}{2}, \frac{3}{2}, \frac{5}{2}, \ldots$) if the nucleus has an odd mass number, it is zero if the nucleus has an even mass and charge number, and a non-fractional integer if the mass number is even and the charge number is odd. Given that the hydrogen ion has a magnetic moment, can we detect its magnetic moment if we take a container filled with hydrogen ions? Probably not, as the vectors point in random directions and cancel each other out (Fig. 10.1).

While the magnitude of $\vec{\mu}$ can be specified, its direction depends on the presence or absence of a magnetic field. To be a bit more accurate, its direction can be specified in the presence of a magnetic field. In this situation, if we define the direction of the magnetic field $\vec{B}_0$ as the z direction, then the z component of the dipole moment aligns with $\vec{B}_0 = B_0\vec{u}_z$. The magnitude of the z component is given by

$$\mu_z = \gamma m_I \bar{h}$$

where m_I is the *magnetic quantum number* which varies from $-I$ to I in steps of 1. In the case of the hydrogen ion $I = \frac{1}{2}$, therefore, m_I can take on a value of $-\frac{1}{2}$ or $\frac{1}{2}$.

The nucleus precesses around the applied magnetic field as a result of the torque expressed by

$$\frac{d\vec{J}}{dt} = \vec{\mu} \times B_o\vec{u}_x$$

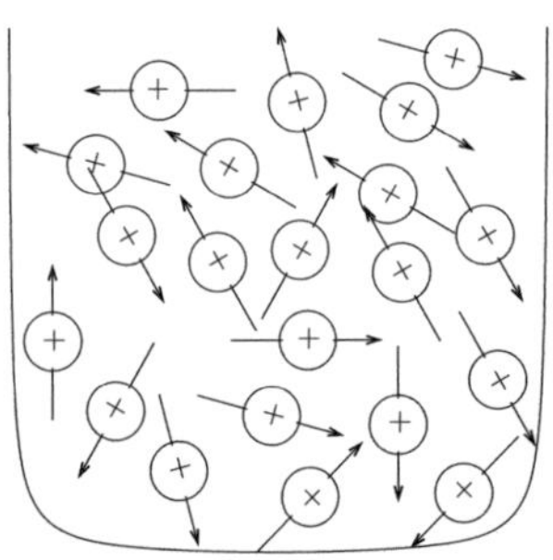

Fig. 10.1 A random collection of protons

or

$$\frac{d\mu_x}{dt} = \gamma B_0 \mu_y$$
$$\frac{d\mu_y}{dt} = -\gamma B_0 \mu_x$$
$$\frac{d\mu_z}{dt} = 0$$

The quantity

$$\omega_0 = \gamma B_0$$

is the resonant frequency and is called the *Larmor frequency*, named after the Irish physicist Joseph Larmor. (You can find a Larmor frequency calculator at https://bio.groups.et.byu.net/LarmourFreqCal.phtml). Solving these equations, we get

$$\mu_x(t) = \mu_x(0)\cos(\omega_0 t) + \mu_y(0)\sin(\omega_0 t)$$
$$\mu_y(t) = -\mu_x(0)\sin(\omega_0 t) + \mu_y(0)\cos(\omega_0 t)$$
$$\mu_z(t) = \mu_z(0)$$

Let's move from individual to collective behavior. Let $\vec{\mu}_i$ denote the moment of the ith nucleus, and let $\vec{M}$ denote the bulk magnetism

$$M = \sum \mu_i$$

Before a magnetic field is applied, as noted earlier, the sum of the moments is zero. After the field is applied, the magnetic moment in the z direction can either be in the positive or negative $\vec{u}_z$ direction depending on the value of m_I. The energy of interaction is given by

$$E = -\vec{\mu} \cdot \vec{B}_0$$

The result of this dot product will change depending on whether the spin is pointing up ($m_I = \frac{1}{2}$) or pointing down ($m_I = -\frac{1}{2}$).

$$E_{up} = -\frac{1}{2}\gamma \bar{h} B_0$$

$$E_{down} = \frac{1}{2}\gamma \bar{h} B_0$$

Because E_{up} is at a lower energy, there will be a slight preference for it, and

$$\Delta E = E_{down} - E_{up} = \gamma \bar{h} B_0$$

will be slightly negative depending on the magnitude of the magnetic field and the temperature of the spin system. The ratio of the number of up-pointing spins and down-pointing

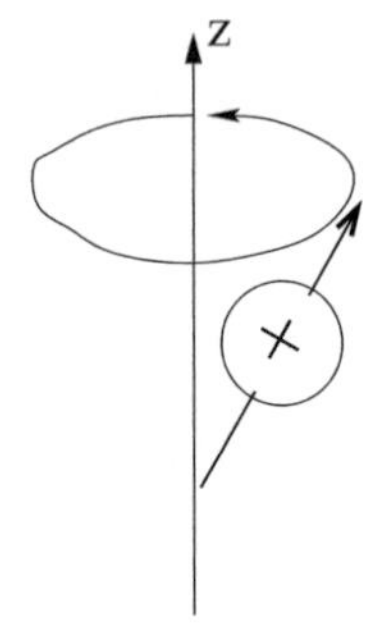

Fig. 10.2 A precessing Hydrogen ion

spins is given by

$$\frac{N_{up}}{N_{down}} = \exp\left(\frac{\Delta E}{kT}\right)$$

As far as the total magnetism is concerned

$$\vec{M} = M_x\vec{u}_x + M_y\vec{u}_y + M_z\vec{u}_z$$

only M_z is nonzero. This is because, as each individual ion precesses around the applied magnetic field, as shown in Fig. 10.2, its x and y components depend on its phase. This phase is randomly distributed; therefore, the sum of the components in the xy plane will be zero. If we take a person and put them in a chamber within a magnetic field as shown in Fig. 10.3, We will generate a small induced magnetic field in the direction opposite to the applied magnetic field. This induced magnetic field will be the sum of the induced magnetic fields due to all the many different ions in the body. This is not particularly useful. For one, this magnetization is static, which makes measuring this field infeasible due to the significant difference in magnitude between the applied magnetic field and the resultant magnetization. Even if it were feasible to measure this field, we couldn't say anything about the body's composition within the MRI machine.

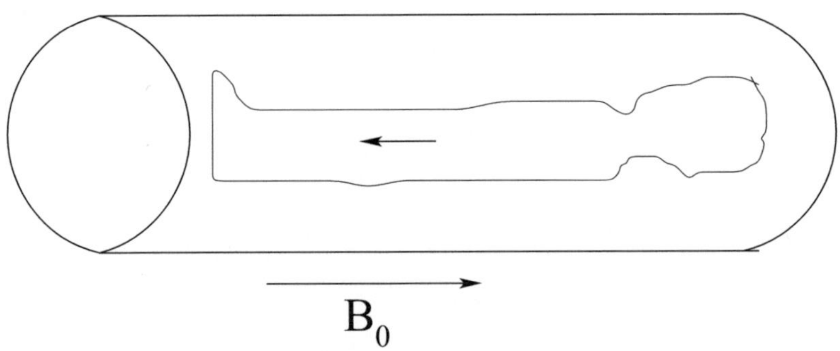

Fig. 10.3 An individual in an MRI machine with only one magnetic field

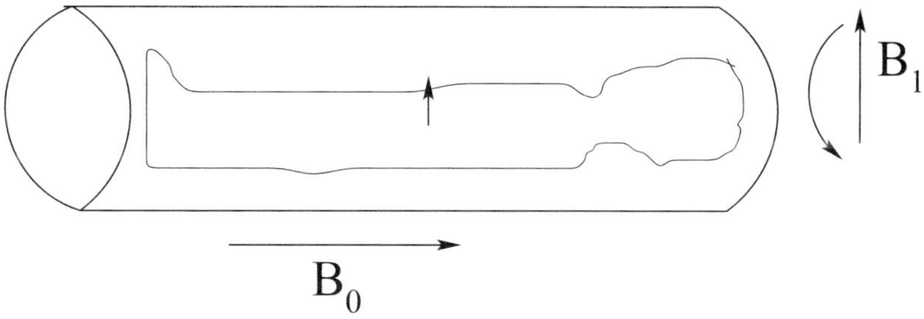

Fig. 10.4 The static magnetic field B_o and the rotating magnetic field B_1

We aim to alter the direction of this magnetization, currently aligned along the z direction, or the longitudinal direction, to the xy plane, or the transverse direction, and we would like to make the magnetization time-varying. This is because it is easy to measure a time-varying magnetic field. All we need to do is place a coil in this time-varying field. A current will be generated in the coil according to Faraday's Law of Induction, which states that a time-varying magnetic field will generate a current in an inductor placed in that magnetic field. The induced current will be proportional to the magnitude of the magnetic field. This allows us to measure the transverse component of the induced magnetic field, which we refer to as the signal intensity of the MRI signal.

We can achieve these objectives if we apply an oscillating field B_1 in the transverse direction, as shown in Fig. 10.4, where the frequency of oscillation is equal to the Larmor frequency ω_o. This will have two effects. The transverse components M_x and M_y will no longer be zero because of increasing coherence between the precessing ions, and the injection of energy will flip some of the up-pointing spins down, resulting in a decrease in the value of M_z. We need to match the frequency of the rotating magnetic field to the Larmor frequency to take advantage of the resonance effect. The resonance effect amplifies the energy injected into the system, allowing us to achieve our objectives with a relatively small (and safer) strength of B_1. Without the resonance effect, the energy required to obtain a measurable transverse component would generate an unacceptable level of heat in the patient. If we leave the oscillating field B_1 on long enough, the longitudinal component will no longer be measurable, and we will only be able to measure the oscillating transverse component.

Let's examine the magnetization in a rotating frame of reference. Initially, before we turn on the rotating magnetic field, we will have all the energy in the longitudinal direction. Ultimately, after the rotating magnetic field, rotating at the Larmor frequency, has been on for a while, we will end up with all the energy in the transverse direction. One way to show this, as depicted in Fig. 10.5, is to use a coordinate system in which one axis is the magnetization in the longitudinal direction and the other axis represents the magnetization in the transverse direction. Note that the transverse component is increasing at the same rate that the longitudinal component is shrinking.

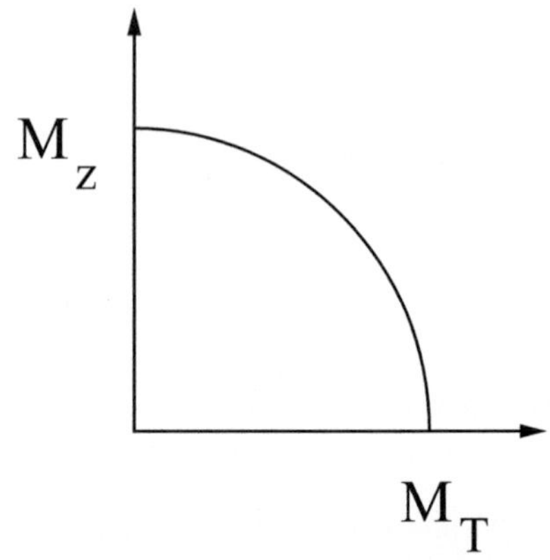

Fig. 10.5 Change in magnetization from the longitudinal direction to the transverse direction

In MRI systems, the range of the applied magnetic field strength is from 0.5 T to 3 T. For Hydrogen ions, this range of applied magnetic field strengths will correspond to Larmor frequencies in the range of 21–128 MHz. The frequencies for FM radio in the US are 88 MHZ to 108 MHZ. Because of this, the turning on and off of the transverse magnetic field is often referred to as the application of a radio frequency (RF) pulse.

When the B_1 field is turned off, both these effects reverse. The relaxation process is given by the *Bloch equations* [91]:

$$\frac{d\vec{M}}{dt} = \gamma \times \vec{B} - \frac{M_x\vec{u}_x + M_y\vec{u}_y}{T_2} - \frac{(M_z - M_z(0))\vec{u}_z}{T_1}$$

where T_1 is called the longitudinal relaxation time or the spin-lattice relaxation time. The latter name comes from the fact that it measures the time required to transfer energy from the spin system, or the spinning nuclei, to the surrounding molecular lattice. T_2 is the spin-spin relaxation time. The spin-spin relaxation is due to the transfer of energy between spinning ions rather than to the lattice. This results in a decrease in phase coherence among the spinning ions and, therefore, a reduction in the transverse component. Rewriting the Bloch equation in terms of its components:

$$\frac{dM_z}{dt} = \frac{(M_z - M_z(0))}{T_1}$$

and the transverse fields decay to zero at the rate

$$\frac{M_x}{dt} = -\frac{M_x}{T2}$$

$$\frac{M_y}{dt} = -\frac{M_y}{T2}$$

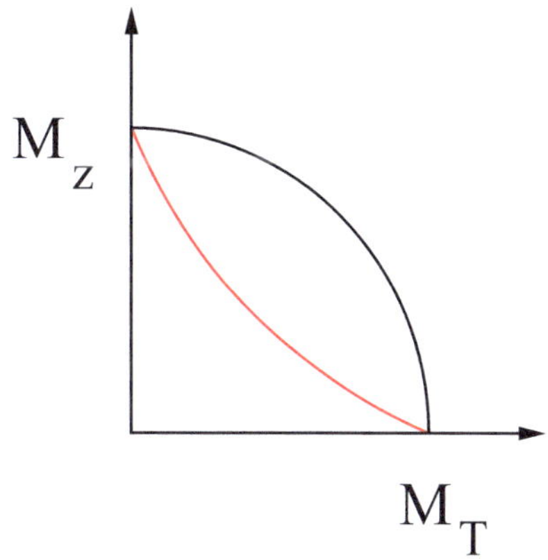

Fig. 10.6 Change in magnetization

T_1 values are mostly affected by the presence of water molecules and therefore provide a measure of the amount of water in the tissues. Because of this

$$M_z \propto N(H)(1 - e^{t/T_1})$$

where $N(H)$ is the number of hydrogen atoms (proton density).

T_2 is the time constant for the decorrelation between the spins of neighboring nuclei and is much smaller than T_1 and is very tissue dependent [92]. Decorrelation between neighboring spins can also occur because of inhomogeneities in the static magnetic field. This decorrelation is not tissue-dependent and therefore has no discriminatory value. Because of the different processes reflected by T_1 and T_2, the rate of change in the longitudinal and transverse components is not the same. Because of this, we observe the situation depicted in Fig. 10.6, where the relaxation curve fails to track the change that occurred when B_1 was turned on.

If we plot the time course of the longitudinal and transverse magnetizations for typical lung tissue values, we obtain the curves shown in Fig. 10.7. The decrease in the transverse component occurs much more rapidly than the increase in the longitudinal component.

10.2 Measuring Signals

After application of the RF pulse, the spin in the xy plane begins to dephase. This dephasing is due to spin-spin interactions with a relaxation time constant T_2 and to perturbations in B_0 caused by material inhomogeneities. For the moment, let's assume that the material being imaged is homogeneous. As mentioned previously, the relaxation time varies for different tissues. For example, in Fig. 10.8 we plotted the transverse components' relaxation times for subcutaneous fat and skeletal muscle. By picking the sampling instant, denoted by T_E (time to echo) appropriately, we can clearly differentiate between subcutaneous fat and skeletal

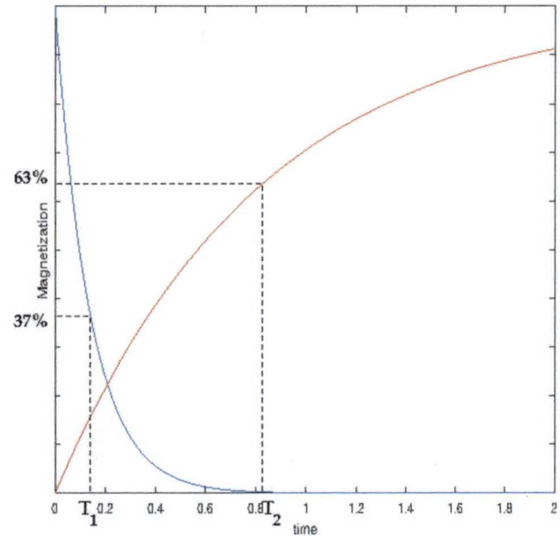

Fig. 10.7 Change in longitudinal and transverse magnetization

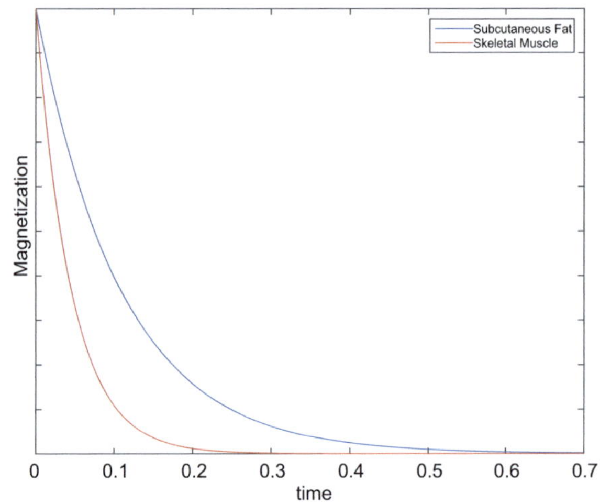

Fig. 10.8 Difference in relaxation times for fat and skeletal muscle

muscle. If we examine the point where the signals differ the most, the appropriate time to sample would likely be later in the interval. However, the longer we wait to sample, the lower the signal values, resulting in a lower signal-to-noise ratio. Therefore, there is a tradeoff between discrimination and signal-to-noise ratio.

Because of the effect of the material inhomogeneities, the time constant of the decay is not T_2 but T_2^*. Effectively, this means that the signal will decay more quickly. Recall that the signal intensity is the magnitude of the transverse component. If the magnitude decays faster, our signal intensity will decrease, giving us a lower signal-to-noise ratio. The spin-spin relaxation is due to the dephasing of the individual spins. Consider the situation in Fig. 10.9. In the figure, we have two vectors, rotating in the clockwise direction, that are initially in phase, and then go out of phase at some time t_o because the red vector rotates faster than the

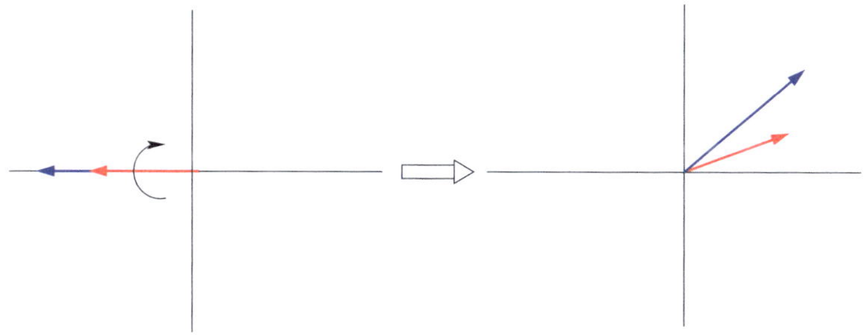

Fig. 10.9 Dephasing of two spins

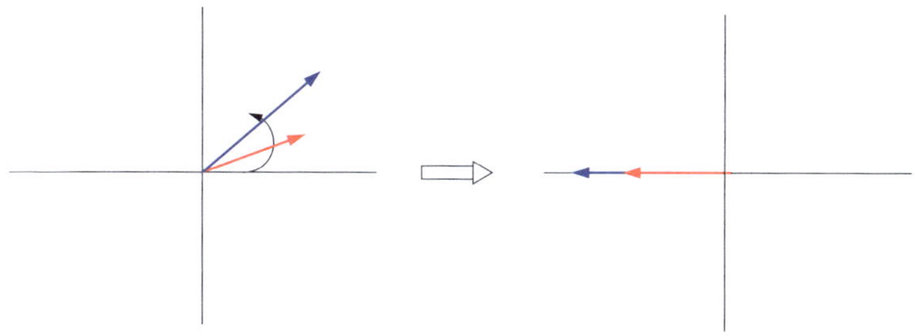

Fig. 10.10 Rephasing of two spins

blue vector. To reverse this dephasing, at time t_o, we apply an input to reverse the direction of the spin. Then, at time $2t_o$, we will have the situation shown in Fig. 10.10. Again, the red vector moves faster than the blue vector, just in the opposite direction, resulting in phase coherence.

This is what we do to recover the value of T_2. If our sampling time is T_E, then at time $T_E/2$ we apply a 180° pulse, which provides rephasing in the opposite direction.

We are showing the decay of magnetization as a smooth exponential; however, that is not entirely accurate. We measure the current generated in our coil by an oscillating magnetic field, so the actual signal we will see is a sinusoid modulated by a decaying exponential as shown in Fig. 10.11. This curve is called a free induction decay (FID) curve. The peak of the FID signal is proportional to the proton density.

So, how do we measure the recovery longitudinal component? The only thing we can measure is samples of the FID signal. The envelope of the FID signal gives us the decay of the magnitude of the transverse signal. Recall from Fig. 10.5 that the maximum value of the transverse component is the maximum value of the longitudinal component at the beginning of the RF pulse. The recovery of the longitudinal signal takes much longer than the decay of the transverse component; $T_1 > T_2$.

We will utilize this fact to obtain an estimate of the longitudinal component. Suppose, after the transverse component has decayed but the longitudinal component has not yet fully

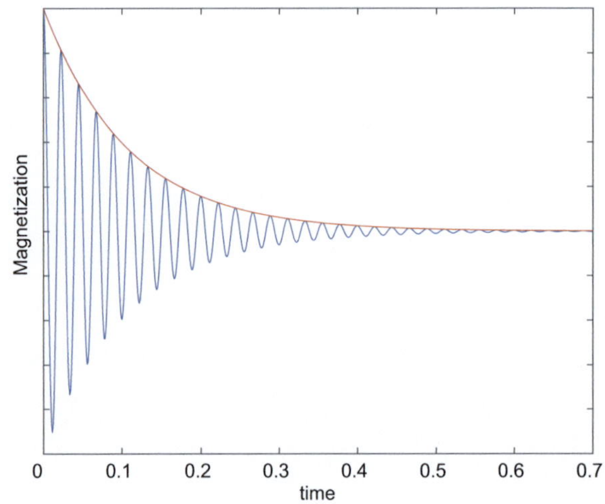

Fig. 10.11 A free induction decay curve

recovered, we apply another RF pulse to the system and measure the transverse component again. This time, the maximum value of the decay curve for the transverse component will be the magnitude of the longitudinal component at the time the second RF pulse was applied. This value is denoted as m_1 in Fig. 10.12. This decrease in the maximum amplitude will be reflected throughout the decay curve for the transverse component, allowing us to infer the value of the relaxation curve for the longitudinal component from the sample of the relaxation curve of the transverse component. If we keep doing this, as shown in Fig. 10.12, we can get a pretty good idea of the longitudinal relaxation curve. Here, T_R denotes the time between repetitions. We repeat the RF pulse every T_R seconds and then sample the transverse relaxation curve T_E seconds after the RF pulse.

Let's summarize. We begin by applying a static magnetic field B_o to the body. Because we cannot measure the resultant induced static magnetic field, which would give us some idea of the type of tissue, we apply a rotating magnetic field B_1, rotating at the Larmor frequency

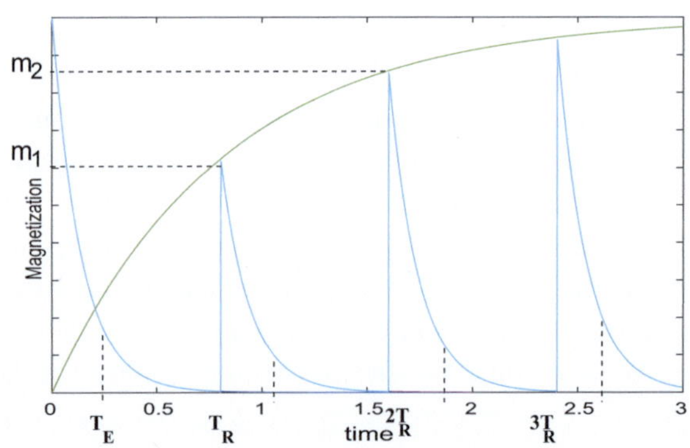

Fig. 10.12 Obtaining the relaxation curve for the longitudinal component

for Hydrogen, which is the resonant frequency for Hydrogen. The rotating magnetic field induces a time-varying magnetization in the body, which is perpendicular to the static field. The time-varying nature of the magnetization enables us to detect the magnitude of this field by utilizing the principle that time-varying magnetic fields induce currents in conducting coils placed within the magnetic field. We turn the rotating magnetic field on and off, generating an RF pulse which causes the transverse magnetic field to increase and then decrease. Because the density of hydrogen ions varies in different tissues, the decay curve will differ accordingly. By sampling the decay curve appropriately, we can distinguish between different tissues. However, nothing we have said so far indicates how we will separate the signals from the different tissues. The signal that we have is the sum of signals from all the different tissues in the body.

10.3 Spatial Localization

Until now, we have only measured the magnetism induced in the body when placed in an external magnetic field. What we want is a three-dimensional image of the body. We will do this in three steps. We begin with the simplest step: measuring the induced magnetism in a slice of the body. Then, we will examine how to transform the slice into an array of voxels or pixels.

10.3.1 Slice Localization

Think of a slice as a flat rectangular surface. This surface can be oriented in various ways. A vector in three-dimensional space can represent this orientation. In the case of MRI this orientation is provided by three sets of coils to induce a gradient in the applied magnetic field in the x, y, and z directions. Recall that we have assigned the z direction to the main applied magnetic field B_0. Let's, for a moment, consider only the gradient fields applied in the positive and negative z directions as shown in Fig. 10.13. Now, instead of the magnetic field having a magnitude of $|B_o|$ along the entire body, the magnitude of the applied magnetic field will vary from $|B_o| - |G_z|$ from one end to $|B_o| + |G_z|$ to the other end of the body. Remember that the energy supplied to the magnetic field to rotate the magnetization into the transverse direction utilizes resonance for energy transfer. Resonance occurs at the Larmor frequency $\gamma|B_o|$, where B_o is the applied magnetic field. In the presence of the gradient fields, the magnitude of the applied magnetic field is B_o at only one point along the z axis. Therefore, in the ideal case, the transverse magnetism will be measured only for that slice of the body experiencing an applied magnetic field of B_o. In practice, the oscillating field B_1 cannot oscillate exactly at the Larmor frequency $\gamma|B_o|$. Instead, the oscillation will contain a narrow range of frequencies around the Larmor frequency. Therefore, the transverse field will reflect a narrow slice through the body. The narrowness depends on the slope of the

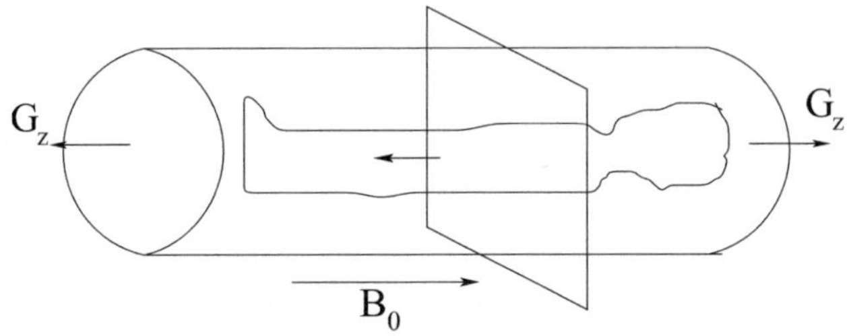

Fig. 10.13 Applying the gradient field in the z direction and selecting a slice

gradient - the steeper the slope, the narrower the slice. By adjusting the gradient fields in the positive and negative z direction, we can move the slice in the corresponding direction. Notice that the selected slice is perpendicular to the net magnetic field. If we now include gradient magnetic fields in the x and y directions, we can orient the applied magnetic field to obtain the desired slice.

10.3.2 Frequency and Phase Encoding

The signal obtained at this point corresponds to the entire slice. Now, let's examine how we can determine the composition of the slice. Let's assume the slice is perpendicular to the z axis for simplicity. Once we have applied the 90° and 180° RF pulses, we can turn off the gradient magnetic fields used for slice selection, and the transverse component of the magnetization will continue to precess at the Larmor frequency in every location in the slice. If we now apply a gradient magnetic field G_x in the x direction, we will linearly alter the frequency of precession along the x dimension of the slice. Looking at Fig. 10.14, all the locations along the green line will be subject to the magnetic field of magnitude $B_o + G_1$, and will precess at the Larmor frequency corresponding to that value, while all the locations along the red line will precess at a frequency corresponding to $B_o + G_2$. This gradient G_x is called the frequency encoding gradient. Note that we still get a single relaxation curve, but the samples we take of the curve will now consist of a sum of frequency components. Each frequency component corresponds to a row of locations along the x axis. How to decompose a sum of frequency components is something we know how to do quite well; we take the Fourier transform. As we operate in the discrete world, this corresponds to the discrete Fourier transform. The number of frequency components depends on the number of samples taken from the FID curve. Each frequency component corresponds to sample values along the x axis.

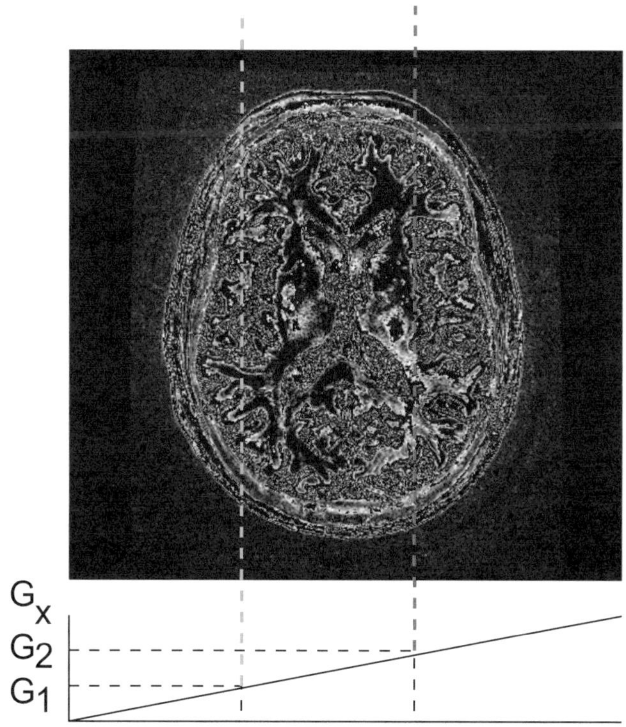

Fig. 10.14 Frequency encoding of the author's brain

We take a slightly different approach to obtain the signal values along the y axis. Consider the situation after we turn off the gradient magnetic fields used for slice selection. If we now turn on and off a linear magnetic gradient along the y direction. This will result in dephasing along the y direction. These phase differences will persist when the frequency encoding gradient is turned on. We note the values of the samples of the decay curve and repeat the whole process with a slightly stronger phase encoding gradient. This will now give us another set of samples. These samples will be signals of the same frequencies as the first set, but with different phases. The rate of change in phase will vary depending on the spatial location. Notice from Fig. 10.15 that the phase changes differ for different locations along the y axis. This is especially true for the steeper gradients. Thus, spatial information is encoded in the phase information. Recall the equivalence between phase shift and time shift. If $X(\omega)$ is the Fourier transform of the signal $x(t)$ then the Fourier transform of $x(t - t_o)$ is $X(\omega)e^{-j\omega t_o}$. We can extend this from a time shift in the one-dimensional case to a spatial shift in the two-dimensional case.

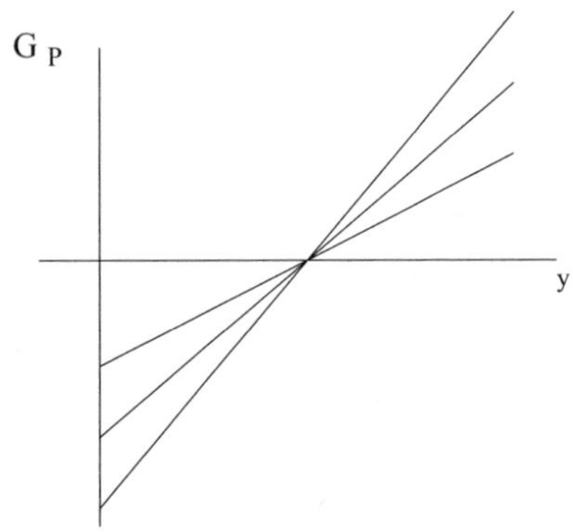

Fig. 10.15 Applying the phase gradient field in the y direction

We continue to repeat this process based on our requirements for resolving the image. If we now take a two-dimensional discrete Fourier transform of this array, it will give us the MR image. Notice that we had to repeat the entire process for each row of our raw data. This means that higher resolution in the y direction is obtained at the cost of greater patient exposure.

10.4 Weightings

We have been discussing sampling the relaxation curve at time T_E as if it were done only once. In practice, we must repeat the process several times at intervals of T_R to obtain a usable signal. The two parameters we can control (apart from the frequency of the RF pulse) are T_R and T_E. We can control these parameters to influence the contrast and relative signal-to-noise ratio (SNR) of the reconstructed MR images. To see how we do this, look at Figs. 10.8 and 10.12. To determine whether we are examining subcutaneous fat or skeletal muscle, we would select a time T_E to sample the relaxation curve. We could choose T_E to be relatively large, allowing for significant contrast between muscle and fat, albeit at the expense of a lower signal-to-noise ratio. We could then wait for the longitudinal component to reach its maximum and send another R_F pulse. This would yield the same relaxation curve for the transverse component, which we could sample again at a time T_E after the RF pulse. If we select T_R large enough for the longitudinal relaxation not to affect our measurement, and T_E is also relatively large, the image we generate is called a T_2-weighted image. If, however, we do not wait until the longitudinal component has completely relaxed, as shown in Fig. 10.12, the starting point of the transverse relaxation curve will be different, resulting in a different value for the sample. The effect of T_1 will thus show up in our sample. The image generated with such a value of T_R is called a T_1 weighted image.

The choice of weighting depends on several factors, including the relative differences between the T_1 and T_2 values of the targets. There is a variation of T_1 and T_2 values depending on the field strengths and among individuals. However, we are more interested in relative

Table 10.1 Approximate T_1 and T_2 values at 1.5 Tesla

	T_1	T_2
Water	4000	2000
Gray matter	900	90
White matter	700	70
Skeletal muscle	1000	50
Fat	250	70

differences, so the individual variations are not that disruptive. In Table 10.1, we show approximate values for T_1 and T_2 in milliseconds, where the field strength $B_o = 1.5T$. In the case of T_2-weighted images, as the relaxation time for water is long, fluid-filled structures appear bright. This makes T_2 weighted images especially useful for studying joints filled with synovial fluid or the ventricles of the brain, which are filled with cerebrospinal fluid. In T_1 weighted images, fat-containing structures show up brightly while fluids show up dark. This makes T_1-weighted images useful when trying to differentiate between situations such as fatty liver disease, where the healthy liver has very little fat compared to the diseased liver.

10.5 Summary

Magnetic resonance imaging is probably the most ingenious of the different techniques we have examined. Using the simple fact that hydrogen nuclei exhibit magnetic resonance, MRI employs various methods to utilize magnetic gradients to extract information. Through signal processing, this information provides a fascinating look inside the human body. We examined how to detect the magnetic resonance of hydrogen nuclei by using a magnetic field rotating at the Larmor frequency of the Hydrogen nuclei. We then examined how to utilize different gradient magnetic fields to extract localization information. This has been a rather quick overview of a complex subject. For those interested in more details there is a very nice set of videos on the subject by Michael Lipton [93].

Fourier Series and Fourier Transform

A signal can be represented as a set of values indexed by time. This is our usual view of a signal. Sometimes the index is continuous, and we get a function $x(t)$. Sometimes the index is discrete, and we get a function $x[n]$. The same signal can also be represented as a set of values indexed by frequency. Again, the frequency index can be continuous or discrete. In this appendix, we examine the continuous-time version of the time and frequency signals; the discrete-time versions are studied in Chap. 2. What allows us to go between the time and frequency representations of the signal is the Fourier transform.

Let's begin with the frequency representation of periodic functions. A periodic function $x(t)$ is one in which

$$x(t) = x(t + kT), \quad k = \pm 1, \pm 2, \pm 3, \dots$$

for all t, where T is the period of the function. In the early eighteenth century, several mathematicians, including Euler, Bernoulli, and d'Alembert, showed that a periodic function can be written in terms of sines and cosines

$$x(t) = a_0 + \sum_{n=1}^{\infty} a_n \cos(n\omega_o t) + \sum_{n=1}^{\infty} b_n \sin(n\omega_o t)$$

where

$$\omega_o = \frac{2\pi}{T}$$

The series representation should probably have been named after Leonhard Euler, who did a lot of the early work in this area, but there were already so many things named after Euler, including Euler's identity, given by

$$e^{j\theta} = \cos(\theta) + j\sin(\theta)$$

K. Sayood, *Biomedical Signal Processing*, Synthesis Lectures on Biomedical Engineering, https://doi.org/10.1007/978-3-032-08855-0

and Euler's number e, (not to be confused with *Euler's constant* or *Euler's numbers*), which is the base of the natural logarithm and also shows up in the preceding equation, that naming the series after him probably seemed like overkill. That is not to take anything away from Joseph Fourier, after whom the series is named, who further developed the idea and applied the expansion to the problem of heat conduction in solids. We can use Euler's identity to write the Fourier series expansion in its *exponential*, as opposed to *trigonometric*, form

$$x(t) = \sum_{n=-\infty}^{\infty} c_n e^{jn\omega_o t}$$

This form is generally more useful in signal processing, so let's explore this a little further. To obtain the expression for c_n, we use the fact that the integral over a period T of the product of $e^{j\omega_o t}$ and $e^{-k\omega_o t}$ is zero except when $j = k$

$$\int_T e^{j\omega_o t} e^{-k\omega_o t} = \begin{cases} 0 & k \neq j \\ T & k = j \end{cases}$$

We can make use of this by multiplying both sides of the exponential Fourier series expression by $e^{-k\omega_o t}$, dividing by T, and integrating over a period. We will end up with

$$c_k = \frac{1}{T} \int_T x(t) e^{-k\omega_o t} dt$$

The sequence $\{c_k\}$ is the frequency domain representation of the signal. That's all well and good, but we generally are not dealing only with periodic functions. What we would like is a way of exploring the frequency content of all kinds of signals, periodic and aperiodic. To get the frequency representation of the aperiodic signals, the exponential form of the Fourier series comes especially handy. Suppose we have an aperiodic signal, as shown in Fig. A.1.

We can generate a periodic signal, $x_p(t)$, from this aperiodic signal, $x(t)$, by repeating $x(t)$ every T seconds (Fig. A.2).

$$x_p(t) = \sum_{n=-\infty}^{\infty} x(t - nT)$$

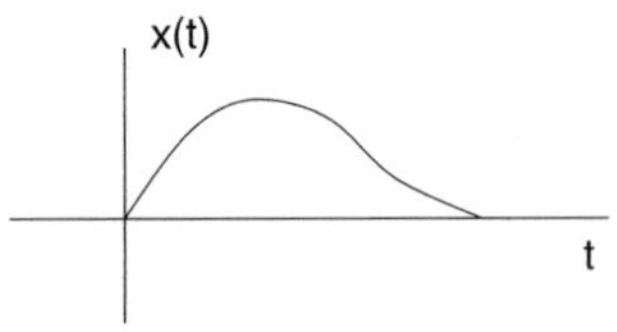

Fig. A.1 An aperiodic signal

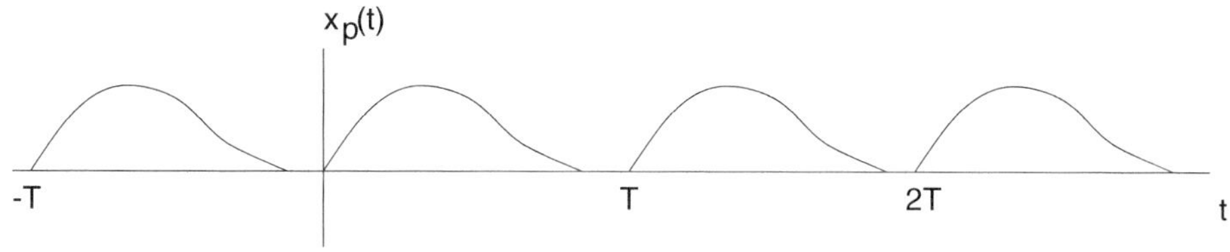

Fig. A.2 Repeating the aperiodic signal to get a periodic signal

To recover the original signal from its periodic extension, all we have to do is take the limit as T goes to infinity.

$$x(t) = \lim_{T\to\infty} x_p(t)$$

To obtain the frequency profile of $x(t)$ we can get the frequency profile of the periodic signal $x_p(t)$ and then take the limit of that as T goes to infinity. When T goes to infinity, the fundamental frequency ω_o goes to zero. So instead of taking the limit as T goes to infinity, we can take the limit as ω_o goes to zero. This is what we will do. To emphasize this, we will replace ω_o with $\Delta\omega$. Writing the Fourier series for $x_p(t)$,

$$x_p(t) = \sum_{n=-\infty}^{\infty} c_n e^{jn\Delta\omega t}$$

$$c_n = \frac{1}{T}\int_{-T/2}^{T/2} x(t)e^{-n\Delta\omega t}dt$$

In the expression for c_n, when we integrate out t, what is left is a function of $n\Delta\omega$. Let's rename $c_n T$ as $X(n\Delta\omega)$ and rewrite the equations above as

$$x_p(t) = \sum_{n=-\infty}^{\infty} \frac{X(n\Delta\omega)}{T} e^{jn\Delta\omega t} = \sum_{n=-\infty}^{\infty} \frac{X(n\Delta\omega)\Delta\omega}{2\pi} e^{jn\Delta\omega t}$$

$$X(n\Delta\omega) = \int_{-T/2}^{T/2} x(t)e^{-n\Delta\omega t}dt$$

Now let's take the limits to recover $x(t)$ and find the frequency profile $X(\omega)$

$$X(\omega) = \lim_{T\to\infty}\int_{-T/2}^{T/2} x(t)e^{-n\Delta\omega t}dt$$

$$= \int_{-\infty}^{\infty} x(t)e^{-j\omega t}dt$$

$$x(t) = \lim_{T\to\infty} x_p(t) = \lim_{T\to\infty,\Delta\omega\to 0} \frac{1}{2\pi} \sum_{n=-\infty}^{\infty} X(n\Delta\omega)e^{jn\Delta\omega t}\Delta\omega$$

$$= \frac{1}{2\pi}\int_{-\infty}^{\infty} X(\omega)e^{j\omega t}d\omega \tag{A.1}$$

$X(\omega)$ is called the Fourier transform of $x(t)$, and $x(t)$ is the inverse Fourier transform of $X(\omega)$.

There are several properties of the Fourier transform that are useful to us. Let's briefly take a look at these.

Time Delay and Phase Change: If the Fourier transform of a time signal $x(t)$ is $X(\omega)$, then the Fourier transform of the delayed signal $x(t-t_o)$ is $X(\omega)e^{-j\omega t_o}$. That is, delay in the time domain corresponds to a phase change in the frequency domain. Showing this is simple. By definition,

$$\mathcal{F}[x(t-t_o)] = \int_{-\infty}^{\infty} x(t-t_o)e^{-j\omega t}dt$$

Substitute $\tau = t - t_o$ in the integral, then $t = \tau + t_o$, and

$$\begin{aligned}\mathcal{F}[x(t-t_o)] &= \int_{-\infty}^{\infty} x(\tau)e^{-j\omega(\tau+t_o)}d\tau \\ &= \int_{-\infty}^{\infty} x(\tau)e^{-j\omega\tau}e^{-j\omega t_o}d\tau \\ &= e^{-j\omega t_o}\int_{-\infty}^{\infty} x(\tau)e^{-j\omega\tau}d\tau \\ &= e^{-j\omega t_o}X(\omega)\end{aligned}$$

Convolution and Multiplication: The convolution operation

$$y(t) = x(t) \circledast h(t) = \int_{-\infty}^{\infty} x(\tau)h(t-\tau)d\tau = \int_{-\infty}^{\infty} x(t-\tau)h(\tau)d\tau$$

provides us with the output $y(t)$ of a linear time invariant system with impulse response $h(t)$ when the input is $x(t)$. In the frequency domain, the output is simply the product of the Fourier transforms of the input and impulse response.

$$\mathcal{F}[x(t) \circledast h(t)] = X(\omega)H(\omega)$$

The proof is straightforward

$$\mathcal{F}[x(t) \circledast h(t)] = \int_{-\infty}^{\infty} \left[\int_{-\infty}^{\infty} x(\tau)h(t-\tau)d\tau \right] e^{-j\omega t} dt$$

Substituting $\sigma = t - \tau$ we get $t = \tau + \sigma$ and

$$\begin{aligned}
\mathcal{F}[x(t) \circledast h(t)] &= \int_{-\infty}^{\infty} \left[\int_{-\infty}^{\infty} x(\tau)h(\sigma)d\tau \right] e^{-j\omega(\tau+\sigma)} d\sigma \\
&= \int_{-\infty}^{\infty} x(\tau)e^{-j\omega\tau} d\tau \int_{-\infty}^{\infty} h(\sigma)e^{-j\omega\sigma} d\sigma \\
&= X(\omega)H(\omega)
\end{aligned}$$

So, convolution in the time domain corresponds to multiplication in the frequency domain. The dual is also true. Multiplication in the time domain corresponds to convolution in the frequency domain.

$$\mathcal{F}[x(t)y(t)] = \frac{1}{2\pi} X(\omega) \circledast Y(\omega).$$

References

1. Nyquist, Harry. 1928. Abridgment of certain topics in telegraph transmission theory. *Journal of the AIEE* 47 (3): 214–217.
2. Shannon, C. E. 1949. Communication in the Presence of Noise. *Proceedings of the IRE* 31: 10–21 January.
3. Cooley, James W, and John W Tukey. 1965. An algorithm for the machine calculation of complex fourier series. *Mathematics of Computation* 19 (90): 297–301.
4. Heideman, Michael T, Don H Johnson, and C Sidney Burrus. 1985. Gauss and the history of the fast fourier transform. *Archive for history of exact sciences*, 265–277.
5. Oppenheim, Alan V, Alan S Willsky, and Syed Hamid Nawab. 1997. *Signals & systems*. Pearson Educación.
6. Sayood, Khalid. 2021. *Signals and systems: a one semester modular course*. Morgan & Claypool Publishers.
7. Signals & systems. 2020. https://www.youtube.com/playlist?list=PLBbeAT-v28f3RoRMG3ZjByxu7c6tlIt_-. Accessed 29 Aug 2025.
8. Butterworth, Stephen, et al. 1930. On the theory of filter amplifiers. *Wireless Engineering* 7 (6): 536–541.
9. McClellan, James H, and Thomas W Parks. 2005. A personal history of the parks-Mcclellan algorithm. *IEEE Signal Processing Magazine* 22 (2): 82–86.
10. Parks, T, and James McClellan. Chebyshev approximation for nonrecursive digital filters with linear phase. *IEEE Transactions on Circuit Theory* 19 (2): 189–194.
11. Evgenij Jakovlevič Remez. 1957. *Obščie vyčislitel'nye metody Čebyševskogo približenija: zadači s linejno vhodjaščimi veščectvennymi parametrami*. Izdatel'stvo akademii nauk ukrainskoj SSR.
12. Lloyd, S. P. 1982. Least squares quantization in PCM. *IEEE Transactions on Information Theory* IT-28: 127–135.
13. Hilbert, E.E. 1977. Cluster compression algorithm-a joint clustering data compression concept. Technical Report JPL Publication 77-43, NASA.
14. Linde, Y., A. Buzo, and R. M. Gray. 1980. An algorithm for vector quantization design. *IEEE Transactions on Communications* COM-28: 84–95.
15. Sayood, K. 2017. *Introduction to Data Compression*, 5th ed. San Francisco: Morgan Kauffman-Elsevier.

K. Sayood, *Biomedical Signal Processing*, Synthesis Lectures on Biomedical Engineering,
https://doi.org/10.1007/978-3-032-08855-0

16. Rosenblatt, Frank. 1958. The perceptron: a probabilistic model for information storage and organization in the brain. *Psychological Review* 65 (6): 386.
17. Vapnik, Vladimir N. 1963. Pattern recognition using generalized portrait method. *Automation and Remote Control* 24 (6): 774–780.
18. Vapnik, Vladimir, and A. J. Lerner. 1963. Generalized portrait method for pattern recognition. *Automation and Remote Control* 24 (6): 774–780.
19. Vapnik, Vladimir N, and Alexey Ya Chervonenkis. 1964. On a class of pattern-recognition learning algorithms. *Automation and Remote Control* 25 (6): 838–845.
20. Boser, Bernhard E, Isabelle M Guyon, and Vladimir N Vapnik. 1992. A training algorithm for optimal margin classifiers. In *Proceedings of the Fifth Annual Workshop on Computational Learning Theory*, 144–152.
21. Cortes, Corinna, and Vladimir Vapnik. 1995. Support-vector networks. *Machine Learning* 20: 273–297.
22. Finkelstein, Gabriel. 2013. *Emil du Bois-Reymond: neuroscience, self, and society in nineteenth-century Germany*. MIT Press.
23. Faber, Donald S, and Alberto E Pereda. 2018. Two forms of electrical transmission between neurons. *Frontiers in Molecular Neuroscience* 11: 427.
24. Lombard, Jonathan. 2014. Once upon a time the cell membranes: 175 years of cell boundary research. *Biology Direct* 9: 1–35.
25. De Palma, Armando, and Germana Pareti. 2011. Bernstein's long path to membrane theory: radical change and conservation in nineteenth-century German electrophysiology. *Journal of the History of the Neurosciences* 20 (4): 306–337.
26. Cole, Kenneth S, and Howard J Curtis. 1939. Electric impedance of the squid giant axon during activity. *The Journal of General Physiology*, 22 (5): 649–670.
27. Katz, Bernhard. 1947. The effect of electrolyte deficiency on the rate of conduction in a single nerve fibre. *The Journal of Physiology* 106 (4): 411.
28. Alan Lloyd Hodgkin. 1939. The relation between conduction velocity and the electrical resistance outside a nerve fibre. *The Journal of Physiology* 94 (4): 560.
29. Hodgkin, Alan L, and Andrew F Huxley. 1952. A quantitative description of membrane current and its application to conduction and excitation in nerve. *The Journal of Physiology*, 117 (4): 500.
30. Chandra, Rohit. 2014. hhrun-hodgkin huxley model simulation for user defined input current. https://www.mathworks.com/matlabcentral/fileexchange/46740-hhrun-hodgkin-huxley-model-simulation-for-user-defined-input-current. Accessed 2024.
31. Lo Jo Mullins. 1959. An analysis of conductance changes in squid axon. *The Journal of General Physiology* 42 (5): 1013–1035.
32. Tasaki, Ichiji, and Susumu Hagiwara. 1957. Demonstration of two stable potential states in the squid giant axon under tetraethylammonium chloride. *The Journal of General Physiology* 40 (6): 859.
33. Hille, Bertil. 1970. Ionic channels in nerve membranes. *Progress in Biophysics and Molecular Biology* 21: 1–32.
34. Hille, Bertil. 1967. *A pharmacological analysis of the ionic channels of nerve*. The Rockefeller University.
35. Narahashi, Toshio, John W Moore, and William R Scott. Tetrodotoxin blockage of sodium conductance increase in lobster giant axons. *The Journal of General Physiology* 47 (5): 965–974.
36. Narahashi, Toshio, and Hans G Haas. Interaction of ddt with the components of lobster nerve membrane conductance. *The Journal of General Physiology* 51 (2): 177–198.
37. Neher, Erwin, and Bert Sakmann. 1976. Single-channel currents recorded from membrane of denervated frog muscle fibres. *Nature* 260 (5554): 799–802.

38. Smith, Christopher Upham Murray. 2012. *The animal spirit doctrine and the origins of neurophysiology*. Oxford University Press.
39. Guyton, Arthur C. 1971. *Human physiology and mechanisms of disease*. W.B. Saunders.
40. Lilly, Leonard S. 2012. *Pathophysiology of heart disease: a collaborative project of medical students and faculty*. Lippincott Williams & Wilkins.
41. Rodney Levick, J. 2013. *An introduction to cardiovascular physiology*. Butterworth-Heinemann.
42. MedRounds 101. 2022. Physiology of st segment elevation and depression explained. https://www.youtube.com/watch?v=_sjmUMmmXEs. Accessed July 2025.
43. Wagner, Galen S, Peter Macfarlane, Hein Wellens, Mark Josephson, Anton Gorgels, David M Mirvis, Olle Pahlm, Borys Surawicz, Paul Kligfield, Rory Childers, et al. 2009. Aha/accf/hrs recommendations for the standardization and interpretation of the electrocardiogram: part vi: acute ischemia/infarction a scientific statement from the American heart association electrocardiography and arrhythmias committee, council on clinical cardiology; the American college of cardiology foundation; and the heart rhythm society endorsed by the international society for computerized electrocardiology. *Journal of the American College of Cardiology*, 53 (11): 1003–1011.
44. Malmivuo, Jaakko, and Robert Plonsey. 1995. *Bioelectromagnetism: principles and applications of bioelectric and biomagnetic fields*. Oxford University Press.
45. Klier, Kristina, Lucas Koch, Lisa Graf, Timo Schinköthe, and Annette Schmidt. 2023. Diagnostic accuracy of single-lead electrocardiograms using the kardia mobile app and the apple watch 4: validation study. *JMIR Cardio* 7: e50701.
46. Sayood, Sinan S, Denice Hodgson-Zingman, and Kevin F Kwaku. 2023. Single limb-lead electrocardiogram accuracy in screening for ventricular pre-excitation. *Circulation* 148 (Suppl_1): A17557–A17557.
47. Sedghamiz, Hooman. 2014. Matlab implementation of pan tompkins ecg qrs detector. *Code Available at the File Exchange Site of MathWorks*, 1–3.
48. MacLean, Paul D. 1990. *The triune brain in evolution: Role in paleocerebral functions*. Springer Science & Business Media.
49. Solms, Mark. 2021. *The hidden spring: A journey to the source of consciousness*. Profile Books.
50. Matthew Cobb. 2020. *The Idea of the Brain: A History: SHORTLISTED FOR THE BAILLIE GIFFORD PRIZE 2020*. Profile Books.
51. Dick, Philip K. 1968. *Do Androids Dream of Electric Sheep?* Doubleday.
52. Clark, Andy. 2013. Whatever next? Predictive brains, situated agents, and the future of cognitive science. *Behavioral and Brain Sciences* 36 (3): 181–204.
53. Buzsáki, György. 2006. *Rhythms of the Brain*. Oxford University Press.
54. Dehaene, Stanislas, and Lionel Naccache. 2001. Towards a cognitive neuroscience of consciousness: basic evidence and a workspace framework. *Cognition* 79 (1–2): 1–37.
55. Mashour, George A, Pieter Roelfsema, Jean-Pierre Changeux, and Stanislas Dehaene. 2020. Conscious processing and the global neuronal workspace hypothesis. *Neuron* 105 (5): 776–798.
56. Baars, Bernard J. 1997. In the theatre of consciousness. Global workspace theory, a rigorous scientific theory of consciousness. *Journal of Consciousness Studies* 4 (4): 292–309.
57. Burrus, C.S., R.A. Gopinath, and H. Guo. 1998. *Introduction to wavelets and wavelet transforms*. Prentice Hall.
58. Daubechies, I. 1992. *Ten lectures on wavelets*. SIAM.
59. Practical Introduction to Multiresolution Analysis. 2020. https://www.mathworks.com/help/wavelet/ug/practical-introduction-to-multiresolution-analysis.html. Accessed 8 April 2025.
60. Shoeb, Ali, Herman Edwards, Jack Connolly, Blaise Bourgeois, S Ted Treves, and John Guttag. Patient-specific seizure onset detection. *Epilepsy & Behavior* 5 (4): 483–498.

61. Trivedi, Priyanka, Vyom Shah, and Santosh Kumar Satapathy. 2023. Recent trends in eeg-based epilepsy detection and classification: a comprehensive review. In *International Conference on Signal, Machines, Automation, and Algorithm*, 409–419. Springer.
62. Mansouri, Amirsalar, Sanjay P Singh, and Khalid Sayood. 2019. Online eeg seizure detection and localization. *Algorithms*, 12 (9): 176.
63. Epilepsy. https://www.who.int/en/news-room/fact-sheets/detail/epilepsy. Accessed 02 Apr 2025.
64. Matis, Georgios, and Theodossios Birbilis. 2008. The glasgow coma scale-a brief review past, present, future. *Acta Neurologica Belgica* 108 (3): 75–89.
65. Ledwidge, Patrick S, and Dennis L Molfese. 2016. Long-term effects of concussion on electrophysiological indices of attention in varsity college athletes: an event-related potential and standardized low-resolution brain electromagnetic tomography approach. *Journal of Neurotrauma*, 33 (23): 2081–2090.
66. Mansouri, Amirsalar, Patrick Ledwidge, Khalid Sayood, and Dennis L Molfese. 2021. A routine electroencephalography monitoring system for automated sports-related concussion detection. *Neurotrauma Reports* 2 (1): 626–638.
67. Berry, Richard B, Rita Brooks, Charlene E Gamaldo, Susan M Harding, Carole Marcus, Bradley V Vaughn, et al. 2012. *The AASM manual for the scoring of sleep and associated events*. American Academy of Sleep Medicine.
68. Solms, Mark. 2000. Dreaming and rem sleep are controlled by different brain mechanisms. *Behavioral and Brain Sciences* 23 (6): 843–850.
69. Steriade, Mircea, Angel Nunez, and Florin Amzica. 1993. A novel slow (< 1 hz) oscillation of neocortical neurons in vivo: depolarizing and hyperpolarizing components. *Journal of Neuroscience* 13 (8): 3252–3265.
70. Steriade, Mircea. 2006. Grouping of brain rhythms in corticothalamic systems. *Neuroscience* 137 (4): 1087–1106.
71. Steriade, Mircea. 2003. The corticothalamic system in sleep. *Frontiers in Bioscience* 8 (4): d878–d899.
72. Lüthi, Anita. 2014. Sleep spindles: where they come from, what they do. *The Neuroscientist* 20 (3): 243–256.
73. Oostenveld, Robert, Pascal Fries, Eric Maris, and Jan-Mathijs. Schoffelen. 2011. Fieldtrip: open source software for advanced analysis of meg, eeg, and invasive electrophysiological data. *Computational Intelligence and Neuroscience* 2011 (1): 156869.
74. Extracting the brain state and events from continuous sleep eeg. https://www.fieldtriptoolbox.org/tutorial/sensor/sleep/. Accessed 289 July 2025.
75. Pitt, Kevin M, Jonathan S Brumberg, Jeremy D Burnison, Jyutika Mehta, and Juhi Kidwai. 2019. Behind the scenes of noninvasive brain–computer interfaces: A review of electroencephalography signals, how they are recorded, and why they matter. *Perspectives of the ASHA Special Interest Groups* 4 (6): 1622–1636.
76. Nirenberg, Lloyd M, John Hanley, and Edwin B Stear. A new approach to prosthetic control: Eeg motor signal tracking with an adaptively designed phase-locked loop. *IEEE Transactions on Biomedical Engineering*, BME-18 (6): 389–398.
77. Tariq, M, PM Trivailo, and M Simic. Eeg-based bci control schemes for lower-limb assistive-robots. *Frontiers in Human Neuroscience* 12.
78. Robinson, Neethu, Ravikiran Mane, Tushar Chouhan, and Cuntai Guan. 2021. Emerging trends in bci-robotics for motor control and rehabilitation. *Current Opinion in Biomedical Engineering* 20: 100354.
79. Vidal, Jacques J. 1977. Real-time detection of brain events in eeg. *Proceedings of the IEEE*, 65 (5): 633–641.

80. McFarland, Dennis J, A Todd Lefkowicz, and Jonathan R Wolpaw. 1997. Design and operation of an eeg-based brain-computer interface with digital signal processing technology. *Behavior Research Methods, Instruments, & Computers* 29 (3): 337–345.
81. Lawrence Ashley Farwell and Emanuel Donchin. 1988. Talking off the top of your head: toward a mental prosthesis utilizing event-related brain potentials. *Electroencephalography and Clinical Neurophysiology* 70 (6): 510–523.
82. Guan, Cuntai, Manoj Thulasidas, and Jiankang Wu. 2004. High performance p300 speller for brain-computer interface. In *IEEE International Workshop on Biomedical Circuits and Systems*, S3–5. IEEE.
83. Fazel-Rezai, Reza, and Kamyar Abhari. 2009. A region-based p300 speller for brain-computer interface. *Canadian Journal of Electrical and Computer Engineering* 34 (3): 81–85.
84. Abdullah Bilal Aygün and Ahmet Reşit Kavsaoğlu. 2022. An innovative p300 speller brain-computer interface design: Easy screen. *Biomedical Signal Processing and Control* 75: 103593.
85. Acqualagna, Laura, and Benjamin Blankertz. 2013. Gaze-independent bci-spelling using rapid serial visual presentation (rsvp). *Clinical Neurophysiology* 124 (5): 901–908.
86. Oken, Barry S, Umut Orhan, Brian Roark, Deniz Erdogmus, Andrew Fowler, Aimee Mooney, Betts Peters, Meghan Miller, and Melanie B Fried-Oken. 2014. Brain–computer interface with language model–electroencephalography fusion for locked-in syndrome. *Neurorehabilitation and Neural Repair* 28 (4): 387–394.
87. Westlake, S, and J Haskel. 2016. How the beatles gave us the ct scan, and what it tells us about innovation. *TheLong+ Short*.
88. Shepp, Lawrence A, and Benjamin F Logan. 1974. The fourier reconstruction of a head section. *IEEE Transactions on Nuclear Science* 21 (3): 21–43.
89. Bracewell, Ronald N. 1956. Strip integration in radio astronomy. *Australian Journal of Physics* 9 (2): 198–217.
90. Landweber, Louis. 1951. An iteration formula for fredholm integral equations of the first kind. *American Journal of Mathematics* 73 (3): 615–624.
91. Bloch, Felix. 1946. Nuclear induction. *Physical Review* 70 (7–8): 460.
92. Bottomley, P. A., C. J. Hardy, R. E. Argersinger, and G. Allen-Moore. 1987. A review of 1h nuclear magnetic resonance relaxation in pathology: are t1 and t2 diagnostic? *Medical Physics* 14 (1): 1–37.
93. Lectures by Dr. Michael Lipton. https://www.youtube.com/playlist?list=PLH_k6d9j2lGUms4QZzokrkXGEoP6oxkYH. Accessed 29 July 2025.

MIX
Papier aus verantwortungsvollen Quellen
Paper from responsible sources
FSC® C105338

If you have any concerns about our products,
you can contact us on
ProductSafety@springernature.com

In case Publisher is established outside the EU,
the EU authorized representative is:
Springer Nature Customer Service Center GmbH
Europaplatz 3, 69115 Heidelberg, Germany

Printed by Libri Plureos GmbH
in Hamburg, Germany